THE CITY

URBAN SOCIOLOGY

SOCIOLOGY OF THE CITY

URBAN SOCIOLOGY

R.N. MORRIS

LONDON AND NEW YORK

First published in 1968

This edition published in 2007
Routledge
2 Park Square, Milton Park, Abingdon, Oxon, OX14 4RN
Simultaneously published in the USA and Canada by Routledge
711 Third Avenue, New York, NY 10017
Routledge is an imprint of Taylor & Francis Group, an informa business

Transferred to Digital Printing 2007
First issued in paperback 2013
© 1968 Routledge

All rights reserved. No part of this book may be reprinted or reproduced or utilized in any form or by any electronic, mechanical, or other means, now known or hereafter invented, including photocopying and recording, or in any information storage or retrieval system, without permission in writing from the publishers.

The publishers have made every effort to contact authors and copyright holders of the works reprinted in the *The City* series. This has not been possible in every case, however, and we would welcome correspondence from those individuals or organisations we have been unable to trace.

These reprints are taken from original copies of each book. In many cases the condition of these originals is not perfect. The publisher has gone to great lengths to ensure the quality of these reprints, but wishes to point out that certain characteristics of the original copies will, of necessity, be apparent in reprints thereof.

British Library Cataloguing in Publication Data
A CIP catalogue record for this book
is available from the British Library

Urban Sociology

ISBN13: 978-0-415-41823-2 (volume hbk)
ISBN13: 978-0-415-41931-4 (subset)
ISBN13: 978-0-415-41318-3 (set)
ISBN13: 978-0-415-85326-2 (volume pbk)
Routledge Library Editions: The City

URBAN SOCIOLOGY

Professor R. N. Morris
The American University, Washington

London
GEORGE ALLEN AND UNWIN LTD
RUSKIN HOUSE · MUSEUM STREET

FIRST PUBLISHED IN 1968
SECOND IMPRESSION 1968
THIRD IMPRESSION 1971

This book is copyright under the Berne Convention. All right are reserved. Apart from any fair dealing for the purposes of private study, research, criticism, or review, as permitted under the Copyright Act, 1956, no part of this publication may be reproduced, stored in a retrieval system, or transmitted, in any form or by any means, electronic, electrical, chemical, mechanical, optical, photocopying, recording or otherwise, without the prior permission of the copyright owner. Enquiries should be addressed to the publishers.

© *George Allen and Unwin Ltd, 1968*

Cloth bound edition ISBN 0 04 300018 5
Paper bound edition ISBN 0 04 300019 3

PRINTED IN GREAT BRITAIN
in 10 point *Plantin* type
BY WILLMER BROTHERS LIMITED
BIRKENHEAD

To Ruth

CONTENTS

ACKNOWLEDGEMENTS	ix
INTRODUCTION	xi
1. 'Urbanism as a Way of Life': The Urban Theory of Louis Wirth	15
2. The Pre-Industrial City	39
3. Wirth's Theory and the Industrial City	62
4. Ecological Processes in the City	101
5. Bureaucracy in Urban-Industrial Societies	114
6. Wirth's Theory of Urbanism: An Overall Evaluation	160
BIBLIOGRAPHY	173
INDEX	174

ACKNOWLEDGEMENTS

It is not easy to recall, let alone to repay, all one's manifold intellectual debts. This is especially true in a work whose aim is to present knowledge in a stimulating and readable form, rather than to report fresh research. The authors from whom I have drawn most will, I hope, be clear from the footnotes.

My original interest in urban sociology was very significantly focused and sharpened by Chelly Halsey, John M. Mogey, and Norman Dennis. More recently, it has been broadened by discussions with my colleagues at American University with John C. Scott, S. Frederick Seymour, and Annabelle B. Motz. To S. Frederick Seymour I owe a particular debt of gratitude. He has most amply and generously placed at my disposal both his time and his encyclopaedic knowledge of the literature. He has devoted many hours to re-reading the manuscript, probing new ideas, and provoking me out of intellectual ruts. His dedication to the development of sociological propositions has had a major impact on this book. His own graduate training at Chicago has provided a valuable perspective on the context of Wirth's thoughts.

Professor W. M. Williams, as Editor of the series, very kindly gave me both warm encouragement and a free hand, along with his wise editorial guidance.

Mark Gordon was an excellent research assistant, spending many hours abstracting materials and ideas from the vast and scattered literature. He also made some very useful comments on the manuscript.

Finally my wife, a fine sociologist in her own right, was both a valuable critic and a major source of moral support, at a time when it was necessary to spend long hours searching for fresh materials.

Obviously, the faults of the book remain my own. I only hope that they will stimulate the reader to make a fuller, more precise analysis of this fascinating subject.

INTRODUCTION

The city has been studied by writers and practitioners from many fields; and diverse points of view can be discerned within any field. Sociology is no exception: sociologists have discussed many features of urban life, utilizing a variety of approaches and stressing different elements. There is no universal agreement among sociologists on which features of the city are central, and which are of minor importance. Indeed, judgments of this nature can be made only as fully-fledged theories of city life are developed, tested and compared. The objective of this book, then, is not to argue that one particular conception of the city is the best; but to examine carefully one conception and one theory, that of Louis Wirth, and to judge its strengths and weaknesses as an aid to understanding urban life. If Wirth's theory is valuable, it will not only give new insights about familiar patterns of social behaviour; it will also raise questions to which no answer is yet available. Theory should be judged by the new questions it provokes, and not merely by the answers it supplies to existing questions.

There have been two major approaches to urban sociology; while few writers have espoused one to the complete exclusion of the other, they point in different directions and ask somewhat different questions. One approach, which we may loosely call the ecological, assumes that the essence of the city lies in the concentration of a very large number of persons in a relatively small space. This approach concentrates on studying the impact of size and density on social organization: the ways in which people differ according to whether they live in a city (containing, say, over 50,000 people), in a town (containing, say, between 2,500 and 49,999 people), in a village (containing, say, between 250 and 2,499 people) or a hamlet (containing less than 250 people). On this definition, a person's status as urban, rural or intermediate is determined principally by the size of the population unit in which he resides; and necessarily changes when he moves from a village to a town. This ecological approach examines the ways in which an increase in the number of residents changes the social organization of the area: its methods of allocating power and reaching decisions, of organizing cooperative activities and restraining conflict, of communicating news and educating the next generation, of maintaining consensus on common objectives and earning a living. Similarly, it is a comparative approach, which seeks features of the city which are relatively constant, not merely from city to city within a society, but preferably from one society to another. It is certainly plausible to expect major differences between the average hamlet-dweller and the average Londoner in many respects; it is less certain that one will find a steady change as one moves from a town of 10,000 to one of 20, then 30, then 40,000; it is unlikely that individual city- or hamlet-dwellers will conform very closely to the average for their category as a whole.

The second approach to urban sociology, which we may loosely call the organizational, begins with patterns of social behaviour rather than with the size of population units. The essence of urban life, on this view, is a particular set of patterns of behaviour, and not the location of one's residence. The key objective, from this viewpoint, is to study the processes through which a particular society encourages or discourages the development of cities, towns, villages and hamlets. Each approach brings together the same two principal elements—a large population unit and a particular set of patterns of behaviour; but whereas the ecological approach tends to treat changes in social organization as the *result* of changes in size, the organizational approach tends to treat them as the *causes* of changes in the size of physical units. On the organizational definition, a person's status as urban, rural or intermediate is determined principally by his behaviour, and does not necessarily change when he moves from a village to a town. As a result, one would expect the behaviour of urbanites to be influenced more by the groups with which they interact than by the fact that they reside in a city or town. It is certainly plausible that cities (or villages) in different societies may be organized very differently; it is less certain that one will be able to arrange each society one observes along a scale which shows the degree to which it is urban; it is unlikely that individuals will exhibit unambiguously all the traits which are said to characterize their society.

While few writers have followed one approach and altogether ignored the other, their divergence is apt to lead to confusion. Part of the confusion arises because sociologists have been too prone to seek very simple accounts of urban life, and to look for a small number of elements from which to expect a full understanding of it. Simplified accounts tend to exaggerate the importance of one or two elements of a complex whole; as a result, their one-sidedness provokes others to build equally simplified counter-theories. One of the significant features of urban development has been the weakening of the ties between a person's place of residence and his group affiliations. The grouping of persons who reside in close proximity no longer coincides so neatly with the social group of persons who are bound by common loyalties and common patterns of behaviour. The degree of coincidence between the physical grouping and the social group may be high or low; but in general it is likely to be lower in the urban areas. In studying urban life, therefore, it is both easy and misleading to concentrate on either the physical grouping or the social group as the one basic element; and in doing so to ignore variations in the extent to which they overlap. The city is significant, not only because it is large and densely settled, not only because its internal organization tends to be different from that of the countryside, but also because it influences and is influenced by groups whose members live far beyond its own boundaries. The city cannot, therefore (if indeed the countryside ever could) be accurately viewed as an independent social system: it is but one constituent part—often a dominant part—of a complex society.

'Urbanism as a Way of Life': The Urban Theory of Louis Wirth

THIS book takes as its starting-point Louis Wirth's classic article, 'Urbanism as a Way of Life', which first appeared in 1938.[1] It was written towards the end of the heyday of the 'Chicago School' in American sociology, and aimed to present a relatively concise theory of city life which would incorporate that School's main research findings over the preceding twenty years. In Wirth's theory, three concepts —size, density and heterogeneity—were taken to be the key features of the city. These key features were then related to each other by a set of propositions, setting out the conditions under which a large, dense, heterogeneous aggregate of people might be expected to cooperate enough to maintain the complex organization of the city. In this manner, the various aspects of city life, as Wirth saw them, could be systematically related to its three key features; and as his propositions were tested, the theory could be confirmed, modified or refuted

This opening chapter presents Wirth's theory, and examines some of the criticisms which have since been levelled against it. There are four sections. The first gives a brief account of the main propositions which Wirth advanced; the second relates some of the warnings which he attached to his analysis of the city; the third section reviews some of the criticisms which have been voiced against his theory. The fourth and longest section considers two omissions in his theory, by discussing some of the classifications of cities, and of the social areas which comprise them.

---

[1] *Amer. J. Sociol.*, vol. 44, 1938, pp. 1–24. It has been reprinted without revision, notably in Hatt, P. K. & Reiss, A. J. (eds.) *Cities and Society*. Free Press, 1957; in Marvick, E. W. & Reiss, A. J. (eds.) *Community Life and Social Policy*. Chicago U.P., 1956; and in Reiss, A. J. (ed.) *Louis Wirth on Cities and Social Life*. Chicago U.P., 1964. Page-references here are to the last-mentioned source.

1. WIRTH'S THEORY OF URBANISM

After the preparatory remarks which justify his definition of the city as a 'relatively large, dense and permanent settlement of socially heterogeneous individuals',[2] Wirth set out to discover the forms of social action and organization that most often emerged in cities, and to show that these could be logically attributed to its unusual size, density and heterogeneity. Wirth did not explicitly argue that these other aspects of city life were directly *caused* by its three key features: he avoided the question of causation by calling them, somewhat vaguely, 'consequences or further characteristics'.[3]

Wirth presented numerous arguments about the consequences for organized social life of large size, high density and heterogeneity. These will be summarized below in the form of twelve propositions;[4] the evidence supporting and modifying some of them will receive attention in later chapters. The order in which the propositions are presented involves small changes from Wirth's original text in an effort to show more clearly the logical connections among them, and to relate each to the achievement of stability in the city.

1. (a) *Growth and diversity are associated in the city with relatively weak bonds among co-residents,* since city-dwellers are less likely than country-dwellers to have lived together for some generations under a common tradition.

(b) (i) *Formal methods of social control must therefore substitute for allegiance to a common tradition;*
and/or

(b) (ii) *The problem of social control in a diverse population will have to be solved by separating the diverse sub-groups physically.* People will tend to react to the great diversity by withdrawing from those with characteristics different from their own. In this manner, relatively homogeneous areas will form within the city; and within these areas stronger bonds may be maintained.

2. (a) *As a town or city grows, it becomes less likely that any resident will know all the others personally;* hence the character of social relationships changes. At the same time, there is an increase in the number of persons whom one meets, and on whom one is to some extent dependent. One is less dependent on particular people; and one's dependence on another is more likely to be limited to a single facet of one's life.

(b) *The majority of one's social contacts in the city are therefore likely to become 'impersonal, superficial, transitory and segmental'.*

---

[2] *Louis Wirth on Cities and Social Life*, p. 66.
[3] *Ibid.*, p. 67.
[4] These have been extracted from *Ibid.*, pp. 69–77.

Each is likely to involve only the surface, or limited segments, of one's personality for a relatively short period. People expect less of those with whom they are in contact in the city.

(c) *The city-dweller is therefore more likely to treat social relationships as means to his own ends,* thus behaving in what Wirth calls a rational, sophisticated manner. In becoming more calculating, he loses some of the 'spontaneous self-expression, the morale, and the sense of participation that comes from living in an integrated society'. Wirth was presumably regarding these as characteristics of primary groups and of rural life in general.

3. (a) *A highly developed division of labour is associated with the emphasis on the treatment of social relationships as means to one's ends.* This is most developed, according to Wirth, in the professions.

(b) *The large firm will tend to dominate the small family business as the division of labour develops.* The large firm has limited liability, can bring together greater resources, and can draw its leaders from a wider circle. Equally important, however, is the fact that, according to Wirth, 'the corporation has no soul'.

(c) *Social integration therefore requires the development of codes of ethics and etiquette for occupational groups*; without such codes, occupational relationships in the city would tend to be 'predatory'.

4. (a) *The elaborate division of labour grows as the market grows.* Not only do cities perform different economic functions from their hinterlands; particular cities will specialize in particular products where this is profitable. In this way, their markets will be national and international, and not merely local.

(b) *Extreme specialization and interdependence is associated with an unstable equilibrium in the city.* Wirth does not explain what he means by this.

5. (a) *As the city grows, it becomes impossible to assemble all its residents in a single place.*

(b) *Therefore increasing reliance has to be placed on indirect communication as a method of spreading information and opinons and of making decisions.* The mass media and the representatives of special interests become important links in the communication between decision-makers and the general public.

6. (a) *As the density of population in an area increases, greater differentiation and specialization tends to result.* This will reinforce the effect of size in inducing greater specialization.

(b) *Greater differentiation and specialization are indeed essential if the area is to support the increased numbers.* This argument is based on the analogy with the survival of animals or plants on a given tract of land.

7. (a) *Physical contacts in the city are close, whereas most social contacts are relatively superficial.*

(b) *People are therefore categorized, and responded to, in terms of visible symbols* such as uniforms and material possessions.

8. (a) *The city's pattern of land use is the result of competition for a scarce resource*: land tends to be owned by those who can derive the greatest economic return from it.

(b) *The desirability of an area for residential purposes is influenced by many social factors*: its prestige, its accessibility to work, the ethnic and racial composition of its population, the absence of such nuisances as smoke, dirt and noise, for example.

(c) *People with similar backgrounds and needs therefore consciously select, unwittingly drift, or are forced by circumstances into the same section of the city.* The different parts of the city thus have different economic functions, and many of them come to be identified with a particular social class or occupational group: Whitehall, the jewellery quarter, Soho, Coronation Street. While these sections may cover a very small area, each one is itself much less diverse than the city as a whole.

9. (a) *The absence of close sentimental and emotional ties between co-workers and between co-residents fosters competition and mutual exploitation rather than cooperation.*

(b) *High population density implies frequent physical contact and living at a fast pace,* since one is constantly receiving new stimuli and meeting new people.

(c) *The combination of frequent contacts and weak emotional ties can be maintained only if there are orderly and meaningful routines* which can control the resultant friction. The traffic light, the policeman, the clock and the conventional working day symbolize these formal routines.

10. (a) *The interaction of persons with very varied roles and personalities breaks down simple class distinctions.* A person belongs to a variety of groups and may be judged by a variety of visible symbols; these may be associated with quite different prestige levels in society.

(b) *As a result, the class structure is less clear;* and a person's positions within it may be somewhat inconsistent with each other.

11. (a) *City dwellers belong to a variety of groups, and their loyalties to these groups often conflict,* since the groups usually call forth quite different aspects of the personality, and since their claims are not necessarily harmonious.

(b) *Consequently the city dweller is more likely to be geographically and socially mobile,* and less restrained by an overriding loyalty to a particular group, home or city.

(c) *Consequently, too, the city dweller is likely to be sophisticated,* in the sense of seeing his loyalties and principles in terms of shades of grey, rather than in black and white.

12. (a) *The division of labour, combined with the emphasis on seg-*

mental relationships, exercises a levelling influence. The symbols, by which the occupant of a role is 'placed' socially, become standardized because this is economically more efficient. Tailor-made articles and specialized or personal services become very expensive, by comparison with those which can be mass-produced for the 'average person'.

(b) *This levelling influence can be seen also in the development of the 'pecuniary nexus'*, the tendency to judge all goods and services against a common standard—money—and to believe that almost any good or service can be obtained if one can only muster enough money to make it profitable.

(c) *This standardization provides elements for a common culture in the society*: both common material and common symbolic cultural objects.

II. THE AIMS OF WIRTH'S THEORY

Before examining the criticisms which have been raised against this theory of urbanism, it is important to outline Wirth's aims, and to clarify the limitations of which he was conscious. In this way, it is possible to dispense fairly quickly with those criticisms which were based on a misunderstanding or mis-reading of his work.[5] It will also be possible in this way to distinguish criticisms which find important gaps in his theory, from those which accuse him of failing to cover matters which he deliberately and explicitly omitted.

Firstly, Wirth was careful not to suggest that urbanism as a way of life was peculiar to city-dwellers; he recognized that the influence of the city stretches far beyond its administrative boundaries, and often across national frontiers. Statistics which show the proportion living in cities of a given size may therefore seriously under-estimate the proportion of persons who are affected by urbanism. Nor did Wirth believe, conversely, that all city-dwellers were fully urbanized in their ways. Many might be recent immigrants from the countryside (or from abroad); and the evidence suggested that the rural way of life rarely disappeared without trace even among the long-term urban residents. While he expected a correlation between residence in an urban area and urban ways, Wirth certainly maintained a clear distinction between the two. Urbanism was to be understood as a set of social institutions and attitudes which would tend to be found whenever people settled permanently in large, dense and heterogeneous groupings. His theory was intended as an attempt to spell out the changes in attitudes and

---

[5] This applies notably to some of the criticisms of Gans, H. 'Urbanism and Suburbanism as Ways of Life: A Re-Evaluation of Definitions', in Rose, A. M. *Human Behavior and Social Processes*. Houghton Mifflin, 1962; reprinted in Rose, P. I. (ed.) *The Study of Society*, Random House, 1966.

social patterns which would result as cities grew indefinitely; to point out the logical implications of city growth for the maintenance of social cohesion. The characteristics of urban life which he outlined were, for him, the logical consequences of accommodating to a large, dense, heterogeneous environment.

Secondly, Wirth recognized that, among cities, size was a poor indication of urbanization: the residents of a small dormitory town on the edge of a conurbation might be more urbanized in their ways than the residents of a larger town in a rural area. He was aware that the extent of urbanization in a town depended not merely on its own size, density and heterogeneity, but also on its susceptibility to the influence of a conurbation. This warning has been carried further by Peter Mann,[6] who used a brief comparison of Forest Row, Sussex, and Huddersfield to argue that the residents of smaller, less dense, more homogeneous towns could be more likely to exhibit those behaviour patterns which Wirth called urban. This contrast underlines a warning about the limitations of the theory, which perhaps deserves fuller emphasis than Wirth gave it: as was stressed in the Introduction, no satisfactory analysis of the city can ignore its position in relation to the surrounding area. Consequently, Wirth's analysis applies to an abstraction—the city which dominates a self-contained metropolitan region—but may apply only poorly to a single town when taken out of its context. The influence of London on Forest Row cannot be ignored in analysing the latter's social structure; similarly an analysis of London would be seriously incomplete if its role in international trade and in Britain's foreign and Commonwealth relations were overlooked. Similarly, it would be a mistake to abstract sections of a city from their context, and to infer that the largest, most dense, most heterogeneous sections of the city were necessarily the most urbanized in terms of their social institutions and the most prevalent attitudes. Wirth's assumption that the influence of a metropolis extends far beyond its administrative boundaries implies that urban ideas and behaviour have already spread over a wide area. It also implies that while the large, dense, heterogeneous main city is crucial in generating urban ideas and behaviour, their diffusion through the hinterland will be influenced by a great variety of factors.

Thirdly, Wirth made it clear that his definition was not intended as a complete description of all the characteristics which cities shared. His objective, rather, was to select the minimum number of key features which—taken in conjunction—were required to account for the nature of urban life. The critic therefore needs to be able to show that some of the important characteristics of urbanization cannot be explained in terms of these three key features; or that one or more of

[6] Mann, P. H. *An Approach to Urban Sociology.* Routledge & Kegan Paul, 1965, pp. 105–106.

these key features is superfluous in the explanation of urban phenomena. Wirth argued that this latter possibility was unlikely, because the city exists when large numbers, high densities and marked heterogeneity are found in *combination*: two of these features would be insufficient to ensure that urbanization took place.

Fourthly, Wirth cautions the reader not to confuse the urban way of life with some of the historical conditions under which it is now found. The development of cities has often coincided with the growth of large scale industrial organization and with modern capitalism; but cities of considerable size have been built in areas with very little capitalistic or industrial development. For a full theory of cities, one needs to be able to separate out the key features of *all* cities from the additional features of cities of a particular type. For this reason, one should be careful *not* to assume in advance (1) that all cities base their economy on the factory and the large firm, (2) that all city centres are surrounded by slums, (3) that mechanical forms of transport are dominant in all cities, or (4) that all cities guarantee their residents a minimum of sanitary services. These features of Western cities have not necessarily been found in cities of other ages or other cultures.

Fifthly, Wirth does not attempt, in his article on urbanism, to classify cities; though he has undertaken this task elsewhere.[7] Obviously in a more detailed analysis of city life one would wish to distinguish those features which are crucial in a particular type of city, but which are not found universally. He mentions this problem in passing, and suggests a classification in terms of location, size, age and function. His classification will be discussed in the last section of this chapter, along with those of other writers. In the present context, he simply notes that there will be differences between, say, an industrial city and a university city, without specifying the nature of these differences.

Finally, Wirth tried to avoid creating the impression that he was devising propositions about cause and effect relationships. He was not asserting that the growths in size, density and heterogeneity were the 'prime movers', and that the other features of the city were merely their consequences. He was not suggesting that size, density and heterogeneity could change independently, and that the other changes in attitudes and social organization would meekly follow them. He was concerned, rather, with deducing the other changes which would need to take place as these key features changed, if the city was to become and remain a stable social unit; to draw out the implications for social organization and social attitudes of changes in these three key variables. In particular, he stressed that, once these other features had come into existence, they might develop or change quite indepen-

---

[7] National Resources Committee, Research Committee on Urbanism. *Our Cities. Their Role in the National Economy*. U.S. Govt. Printing Office, 1937, p. 8.

dently of changes in the key features which they had originally accompanied.

III. THE MAJOR CRITICISMS OF WIRTH'S THEORY

The criticisms which have been levelled against Wirth's theory are numerous. If those which are based on misunderstandings are omitted —such as the criticism that not every relationship in the city is secondary, and not every relationship outside the city limits is primary—the remaining objections can be arranged into four groups.

The first criticism is that, although Wirth desired generalizations which would hold for all cities, some of his deductions apply only to industrial cities, if at all. This objection applies most strongly to a later section of Wirth's article, dealing in much more detail with the impact of urbanization on the family, and applying to this situation the propositions which were laid out earlier. It also has relevance for the general propositions, however. Wirth's own experience, and the research upon which he could draw, was mainly based on American cities in the twenties and thirties—a period when American society, and particularly its cities, were recovering from the impact of the mass immigration from much less developed countries in the early years of the century. This was the period of the Great Depression and Al Capone; a period during which the inner areas of Chicago had received unusual attention from sociologists, but the more outlying areas had been relatively unstudied. Wirth's critics have often maintained that his account of the city placed too much emphasis upon its problems and disorganization. More crucially in the present context, they have argued that some of his propositions hardly apply to the pre-industrial city: this criticism would seem to relate especially to propositions 3b, 8a (concerning the emphasis on land use in terms of economic profitability), 9a (concerning co-workers), 10a, and possibly 12. The evidence which has accumulated since Wirth wrote, and which will be discussed later, has revealed that the pre-industrial city is not dominated by large firms which have no souls, although the instruments of government may be large and complex. The relationship between the use to which a piece of land is put and its economic profitability is not always apparent in the pre-industrial city; and indeed there are conspicuous exceptions such as parks in the industrial cities of the West. Relations between co-workers in the pre-industrial city are often closer to those in an 'old-fashioned' family business than to those in a large, competitive company. Interaction between members of the different classes in the pre-industrial city may be too limited to blur class distinctions: indeed, Sjoberg[8] concludes that the gulf between the classes is particularly clear-cut. Finally, the survival of

---

[8] Sjoberg, G. *The Pre-Industrial City*. Free Press, 1960, chaps. I, VI.

small businesses in the pre-industrial city means that the levelling and unifying influences associated with mass production are much less apparent than in the industrial city. While goods and services are frequently judged in monetary terms, most goods and services in the pre-industrial city remain unstandardized—indeed, the weighing and measuring instruments, which symbolize standardization in the West, may be subject to wide fluctuations. This general line of criticism implies, of course, that Wirth's theory needs modification; but certainly does not suggest fundamental inadequacies.

The second objection concerns Wirth's view that relationships in a rural society tend to be primary, while those in an urban society tend to be secondary in character. Again, the objection takes several related forms. Sociologists who have studied rural areas in Britain, for example, have sometimes found social relationships to be complex, in the sense that two villagers might know each other in a variety of roles: as neighbours, as relatives by marriage, as members of different churches, as members of the same age-set, and as patrons of different social clubs. They have pointed out that this complexity does not mean, as Wirth tended to imply, that the two persons' overall relationship will be warm and primary; although it does follow that each has much more to lose if he should antagonize the other, and correspondingly much more to gain if the two cooperate. Mann[9] suggested that the change from rural to urban society did not involve a *replacement* of primary by secondary relationships, with the result that the number of primary relationships one made in an urbanized area was smaller; but simply that the number of secondary relationships one establishes rises faster than the number of primary relationships as one moves from a rural to an urbanized area, with the result that only the *proportion* of primary relationships in the city is smaller. Where the objection is levelled against the analysis of urban rather than rural life, it takes a slightly different form. Some societies are so urbanized that very few significant urban-rural differences remain; in consequence, the differences in density, heterogeneity and size are not reflected in cultural differences.[10] Oscar Lewis[11] went further, and argued that there might be no single continuum on which all rural and urban societies could be placed.

The third and fourth criticisms are related; in some respects they are the most important for a full theory of the city. The third criticism which has been raised is that the very variety within the city makes many of Wirth's propositions over-simplified. Since he has defined the

[9] *Op. cit.*, pp. 99–100.
[10] Some of the criticisms of Gans, *op. cit.*, pp. 307–315, are relevant here.
[11] Lewis, O. 'Further Observations on the Folk-Urban Continuum and Urbanization with Special Reference to Mexico City', in Hauser, P. M. & Schnore, L. F. (eds.). *The Study of Urbanization*. Wiley, 1965, pp. 497 *et seq.*

city, in part, as an unusually heterogeneous grouping, Wirth errs in expecting to be able to make simple propositions about its behaviour patterns. Mann, for instance, has contended that the main feature of urban social relationships is their variety rather than their impersonality.[12] This is not simply a re-phrasing of the argument that not all city-dwellers exhibit urban characteristics and behaviour. It represents in part a realization that one is omitting very significant aspects of city behaviour if one concentrates on its impersonal features. The urban resident has far more opportunities than the rural resident to form secondary relationships; but he may also have more opportunities to form primary relationships with people of his own choosing. There are greater opportunities for him to avoid church-going without detection or stigma; but there are also greater opportunities for him to choose a church which shares his particular beliefs. He is more likely to remain friendless in the city; but he also has a wider range of casual acquaintances from whom to choose his friends. While certain forms of social life flourish in the city and others struggle, the city usually tolerates a much broader range of groups and attitudes. Oscar Lewis[13] is the clearest advocate of this criticism; he argues that the city as a form of social organization differs from the rural area in two key respects: it offers a greater range of alternatives, especially if it is large; and the groups within the city differ widely in the extent to which they can and do partake of this wider range. The city-dweller may choose to be cosmopolitan or parochial in his outlook and tastes; both outlooks are found in the city, and it would be misleading to label one urbanized and the other (by default) rural.

The fourth and final criticism of Wirth's theory is related to the third. Wirth describes some of the conditions which must be met if a large, dense, heterogeneous grouping of people is to be stable and to operate without excessive friction; these 'further characteristics' are clearly outlined. Yet he does not go on to examine the *consequences* of these further characteristics, or the conditions under which they in turn may cause instability. Consequently, according to his critics, he misses out important stages in the process whereby urban characteristics influence personality and attitudes; and at times these omissions lead him to erroneous conclusions. His studies of the ghetto[14] made him well aware that many small sectors of Chicago were inhabited by relatively homogeneous groups, which shared a common cultural background and a strong social life, but whose contact with the remainder of the city was relatively restricted and formal. Yet these enclaves receive little attention in his theory. In part, these studies led to un-

---

[12] *Op. cit.*, p. 103. He does not elucidate the difference, if any, between his term 'variety' and Wirth's 'heterogeneity'.
[13] *Op. cit.*, pp. 497 *et seq.*
[14] Notably Wirth, L. *The Ghetto*. Chicago U.P., 1928.

warranted conclusions, because they concentrated on the enclaves of recent immigrants from abroad. It was frequently found that the immigrants' children adopted American speech and behaviour, and moved out of their parents' sector of the city as they matured; until the late twenties, when the restrictive immigration laws began to take effect, they were quickly replaced by other recent immigrants from the same country. The function of these ghettoes in familiarizing immigrants with American ways was much more conspicuous than the fact that some immigrants remained in them; and tended to obscure the existence of similar enclaves of working-class native whites who were fairly stable residentially. Today it is much easier to see that areas of this nature are an integral part of cities, and not merely temporary training-centres for the recent arrivals. A fuller theory of urban life, then, needs to recognize that primary groups are an integral part of urban society, and perform significant functions: they are not merely remnants of rural life, but intrinsic elements in all known societies. A fuller theory of urban life also needs to show the ways in which primary and secondary groups are interdependent in an urban setting; and what kinds of primary groups are necessary to offset the impersonality of more formal relationships. Sociologists are well aware that one never finds a perfect example of a primary or a secondary group: yet it is very difficult to find in the literature a convincing explanation for this apparently universal law.

The task of the present section has been to present but not to evaluate these criticisms; although the trivial ones have deliberately been omitted. Judgment of their merit, and of the modifications which should be made to accommodate them, will be postponed until the concluding chapter, when some of the evidence has been presented. The concluding section of this chapter turns to one topic which Wirth mentions but does not integrate into the body of his theory: the problem of classifying cities and areas within a city, in terms which are consonant with the main theory. The attempt to do this should illuminate our assessment of the usefulness of the theory.

IV. CLASSIFICATORY SCHEMES FOR CITIES AND THEIR COMPONENTS

(a) Classifications of Cities

Before discussing the schemes of Wirth and others for classifying cities, it may be useful to relate such schemes to urban theory. The relationship between a classificatory scheme and a theory is two-fold. Firstly, the classificatory scheme is based on concepts, which are the elements from which theories are made. To be useful, a concept must be precise and communicable; it must also point to significant similarities and differences among the objects being classified. The scheme used for classifying books in a library is an obvious example. To be useful, it must be precise enough so that each book has one correct location on the shelves; and if possible, the location number of a book should be the same in most libraries. It must also group the books meaningfully, so that one may find books on the same topic all in the same place. At the same time, it is clear that such a scheme for classification is not a theory: a theory requires not merely concepts, although these are an essential ingredient; it also requires that the concepts be related to each other to yield propositions which can be tested against the facts. Continuing the example of library classifications, no clear theory would exist until testable propositions were available for grouping the books on different topics. As examples of such propositions, one might propose: (i) sociology and economics books are closely related, because sociology students' outside reading is more likely to be in economics than in any other field; (ii) books on a particular composer should be placed alongside records of that composer's works, rather than with books on other composers, because people who wish to borrow the records are more likely to use these books than anyone else. These propositions linking the concepts into a consistent and testable whole form the second essential ingredient of theory. A classificatory scheme without propositions is not, in this sense, a theory at all; though it contains the necessary concepts.

Many classificatory schemes have been proposed for cities; some of them have been simple schemes using only one type of variable. The simplest classification would relate to size alone; this has very rarely been used by sociologists, except to reveal urban-rural differences. Mann[15] used the British 1951 Census materials to see how many of the traditional urban-rural differences were still apparent. In much more detail, Duncan and Reiss[16] divided American cities into eleven size groups to see which of the variables measured by the 1950 Census changed regularly as city size increased. A few researchers have pointed to a single social or cultural variable. Redfield and Singer,[17] for example, distinguished the orthogenetic from the heterogenetic city, where the former functioned to justify and strengthen the existing culture and institutions, while the latter functioned as a stimulus to social change. Weber[18] separated patrician from plebeian cities, in terms of the social class which had most political power: and in Hauser and Schnore[19] the distinction is made between primitive and feudal cities—the latter having a written language and a well-defined literate

[15] *Op. cit.*, pp. 30–65, 69–70.
[16] Duncan, O. D. and Reiss, A. J. *Social Characteristics of Urban and Rural Communities, 1950.* Wiley, 1956.
[17] Redfield, R. and Singer, M. B. 'The Cultural Role of Cities'. *Econ. Devel. & Cultural Change*, vol. 3, 1954.
[18] Weber, M. *The City.* (Eng. trans.) Free Press, 1960.
[19] Hauser and Schnore, *op. cit.*, p. 171.

elite, and hence being presumed more able to resist the impact of industrial urbanization on their social structures. Finally, cities in Southern Africa have been divided into colonial and pre-colonial,[20] where the latter existed as urban areas before European colonization. Mediaeval cities have been divided into those with an economic orientation and those with a political or intellectual orientation.

The remaining simple classificatory schemes have concentrated on economic variables. Breese[21] suggested a classification of cities as industrial, administrative, colonial or market; while Lampard[22] asserted that in the nineteenth century, the particular dominant industry was a useful basis for classification, since this was closely related to the rate of economic growth. Harris and Ullman[23] used the categories central-place cities, transport cities and specialized function cities. Central-place cities are those whose influence extends over a wide hinterland which includes a variety of cities with minor influence. Transport cities are those where goods frequently change from one transport line, or even one mode of transportation, to another. Karl Marx categorized cities in terms of the relations of production, distinguishing slave-owning, feudal, capitalist and socialist cities.[24] Vance and Sutker, and Bogue,[25] classified them in terms of their share in the organization and financing of trade, which they took as a measure of the size of the city's hinterland and its dominance over that hinterland. Finally, Hoselitz[26] pointed to the city's function in the economic growth of the region or country: generative cities were those which made a positive overall contribution to the economic development of the area, while parasitic cities tended to consume more than they produced and thus restricted the area's economic growth.

All these classificatory schemes suffer from a common defect: they are too simple. Each assumes that one element or one social institution dominates the city, and is not only a necessary but also a sufficient condition for a full description of the city. It is clear that university towns, coal-mining towns and ports have distinctive features which are related to the differences in employment opportunities which they offer;

[20] United Nations. *Report on the World Social Situation*. U.N., 1957.
[21] Breese, G. E. *Urbanization in Newly Developing Countries*. Prentice-Hall, 1966, p. 3.
[22] Lampard, E. E. 'The History of Cities in Economically Advanced Areas', *Econ. Develop. & Cultural Change*, vol. 3, 1955, pp. 107–113.
[23] Harris, C. D. and Ullman, E. L. 'The Nature of Cities'. *Ann. Amer. Acad. Polit. Soc. Sci.*, No. 242, 1945.
[24] Quoted in Hauser and Schnore, *op. cit.*, p. 168.
[25] Vance, R. B. and Sutker, S. S. 'Metropolitan Dominance and Integration', in Vance, R. B. and Demerath, N. J. (eds.) *The Urban South*. The North Carolina U.P., 1954, pp. 124–128; Bogue, D. J. *The Structure of the Metropolitan Community*. U. Michigan P., 1949.
[26] Hoselitz, B. F. 'Generative and Parasitic Cities'. *Econ. Develop. & Cultural Change*, vol. 3, 1955.

and that developments in the major industry affect not only its employees but also all those families who provide them with goods and services. Nevertheless, the dominant industry does not influence all groups in the city in the same direction or to the same extent; nor are conditions in an industry uniform throughout a country, much less throughout the world. More important, relatively few towns can be accurately described in precise terms. To classify a town as 'manufacturing' is not a very exact description: to refer to its economy as mixed or balanced is even vaguer. This is a crucial objection in a complex economy, where one can no longer say that Oxford (for example) is purely a university town, Sheffield purely a steel town, or Stoke-on-Trent purely a pottery-making town.[27] As the size and complexity of a town increases, the probability of finding a category which is both precise and accurate decreases rapidly.

This problem has been partly met by more complex classifications, such as that of Nelson.[28] These recognize the possibility that cities may be unusually well supplied with workers in several industries or occupational groups; though basically they continue to stress only one underlying type of variable.

The alternative approach is to devise a set of categories based on two or more variables at a time. The broadest and most widely used distinction, which was most fully described by Sjoberg,[29] is between the pre-industrial and the industrial. This distinction has been made and elaborated by many writers; although Jones has argued that some writers have used the former as an undifferentiated residual category.[30] Wirth, for example, held the view that the coming of industrialization and capitalism changed the city radically, because it led to the growth of giant organizations.

Sjoberg divided societies into three types: the folk, the feudal and the industrial-urban. The folk society, he argued, is small, self-sufficient and pre-literate; lacking any real division of labour or complex technology, it could not give rise to towns or cities. The feudal society, in which the pre-industrial city is found, is literate, and sufficiently advanced to have an agricultural surplus. Although it still depends on human and animal sources of energy, it has a leisured upper class who live in cities. The cities are larger, denser, more heterogeneous and contain a wider range of non-agricultural specialists than the villages. There is a clear class structure, and an oligarchy

[27] Indeed, these cities are classified as follows in Moser, C. A. and Scott, W. *British Towns*. Oliver and Boyd, 1961: Oxford—Mainly Professional, Spa and Administrative Centres; Sheffield—Mainly Railway Centres; Stoke-on-Trent—More Recent Metal Manufacturing.
[28] Nelson, H. J. 'A Service Classification of American Cities'. *Econ. Geog.*, vol. 31, 1955.
[29] *Op. cit.*, esp. chaps. I, XI.
[30] Jones, E. *Towns and Cities*, O.U.P., 1966, p. 39.

who hold religious, educational and political power and who tend to shun economic activity. The industrial-urban society, in which of course the industrial city is found, has a far higher proportion living in cities, much more technical knowledge, and uses inanimate sources of energy. Its class system is more likely to be based on achievement in large-scale organizations; science and mass education are important, and norms are permissive rather than prescriptive. Obviously not all societies can be fitted neatly into one of these three categories; and not all cities in a given society need be of the same type; again we are dealing with abstractions, attempts to draw out in clear and over-simplified form the main characteristics of a society and its cities.

These two categories, the pre-industrial and the industrial city, have been fairly widely accepted as basic types: one can find examples of each, and cases where a city appears to be making the transition from pre-industrial to industrial. A number of authors have sub-divided each type to improve the classificatory scheme. Hauser[31] described two types of industrial city, which he called the industrial and the metropolitan; in the former, activities tend to be strongly drawn towards the city centre and land prices fall rapidly as one moves outwards. In the metropolitan city, which is significant in the world as well as in the regional economy, activities tend to move away from the centre towards the suburban area; and the relationship between land prices and distance from the city centre is much less clear. Wirth's own classification, which was designed mainly for industrial cities, used a classification in terms of six variables: these explicitly included size, location relative to other cities, economic function, and age. In his more elaborate analysis, he also referred to their economies as balanced or unbalanced, and as prosperous or declining; these constitute the fifth and sixth variables in his scheme.[32] It is interesting to note that Wirth included directly only one of the three key features—size. Heterogeneity may be implicit in the scheme, in so far as it is related to economic function; though Wirth was imprecise about the respects in which heterogeneity was important. Density enters only indirectly, if at all, through its potential connections with age and prosperity. For the most part, density and heterogeneity were seen as necessary qualifications for a city, but not as distinguishing marks for the various types of city.

The final type of classification of industrial cities is that which emerges from an elaborate multi-factorial analysis of census data. The most ambitious ventures have been those of Moser and Scott in England, and Hadden and Borgatta in the US[33] These studies, which

[31] Hauser and Schnore, *op. cit.*, p. 4.
[32] National Resources Committee, *op. cit.*, p. 8.
[33] Moser and Scott, *op. cit.*, Hadden, J. K. and Borgatta, E. F. *American Cities.* Rand McNally, 1965.

will be considered in detail in Chapter 3, have essentially analysed all available census data on each city within a country, with no prior selection in terms of theoretical relevance. The resulting factors, as the authors recognize, are strongly influenced by the range of available official statistics.

It is difficult to integrate these empirical classifications into conventional sociological theory. On the one hand, the data they have assembled from official sources is detailed and valuable; on the other, their approach was not designed to test theory, and a re-analysis would be necessary to produce a more appropriate set of groups for purposes of theory-testing. Probably their most important contribution to sociological theory was their emphasis on social class as a variable to use in categorizing towns. It was not clear how far this was independent of differences in economic function, which Moser and Scott did not study systematically; but even if the two are closely related, social class may be the easier to use in future analyses.

Turning to pre-industrial cities, Redfield and Singer[34] divided them into two types: the administrative and cultural cities, which included those established or developed mainly by colonial governments; and the native commercial cities, which were flourishing before colonial days. The Japanese studies reported in Hauser and Schnore[35] identified five types of city, three of which might be described as pre-industrial: provincial capitals, temple or shrine towns, and postroad stages.

The most careful classification of pre-industrial cities was that of Max Weber, who found three types within the context of European and South European mediaeval.[36] Weber pointed to eleven bases for distinguishing among these three types: the location of the patricians' castles; the dependence of the city budget on civic gratuities and rent distributions; the fear of reform movements initiated by the children of debtors; attitudes towards slave labour; identification of the clan with the territorial areas; the development of guilds; the social stratum on which democracy was founded; the extent of orientation towards the interest of the producers in city politics; dominance of the city over its hinterland; the existence of status differences in the cities between the freemen and the freedmen; and the interest in exploiting conquered land for production and trade, or for immediate consumption and destruction.

Two final classificatory schemes relate to stages in the *process* of urbanization, rather than to types of city in a static sense. Lampard[37]

[34] *Op. cit.*, pp. 56–57.
[35] *Op. cit.*, esp. pp. 322–323.
[36] *Op. cit.*, pp. 197–223.
[37] Lampard, E. E. 'Historical Aspects of Urbanization', in Hauser and Schnore, *op. cit.*, pp. 528–547.

differentiates four stages in the process of urbanization: the primordial, the definitive, the classic, and the industrial. In the primordial stage, there is for the first time an economic surplus and urban organization becomes possible as long as the town remains close to its food supplies. In the definitive stage, the development has proceeded to the point where it is possible to found and transplant towns; but the towns are still closely integrated with their immediate hinterlands. The classic stage sees a weakening of the links between the town and its immediate environment, but agricultural production is not yet specialized, and most trade is still local. The industrial stage witnesses the removal of many of the restraints on city growth; some constraints remain, but there are no longer obvious limits to the proportion of the population which may become urban. Reissman's[38] scheme is perhaps the most ambitious attempt to present a consistent analysis of urbanization. He argues that large-scale industrialization requires an elaborate technology and well-developed physical sciences, and that these place definite limits on the variety of social institutions which can accompany industrial urbanization.[39] While there are differences of detail from one country to another, Reissman sees the process of industrial urbanization essentially as a single process with four main constituents: growth in the proportion of the population who live in cities; development of large-scale manufacturing and the large-scale organization of private and business services; the emergence of a middle-class of entrepreneurs and professionals who are committed to social change and who have much to gain from industrial urbanization; and nationalism or a similar ideology which stresses the common interest of all residents in economic development and material prosperity, and which opposes local loyalties and sectional traditions. These four constituents together are seen as conditions for industrial urbanization: success in the transition from a feudal to an urban-industrial society is assumed to depend on balanced (though not necessarily delicately balanced) simultaneous development in all four respects. Regrettably, Reissman lacks the data to test his argument carefully; and some of his thumbnail sketches of countries undergoing industrial urbanization are misleading or inaccurate, which casts doubt on an otherwise plausible argument. It also seems more reasonable to regard Reissman's four constituent elements as necessary rather than as sufficient conditions for industrial urban growth. The classificatory scheme which was developed from these elements contained 256 categories; on each of the four constituents, a country might be

[38] Reissman, L. *The Urban Process*. Free Press, 1964, esp. pp. 212–234.

[39] This does not necessarily mean, of course, that industrialization is reducing the differences between societies and their cities. See the discussion on this point in Halmos, P. (ed.) *The Development of Industrial Societies*. Sociological Review Monograph No. 8. Keele, 1964.

placed in the top 25%, the second 25%, the third 25%, or the bottom 25%. This scheme has two strengths: it can make relatively fine distinctions between countries, and it reveals how closely the four constituent elements tend to vary together; one of its drawbacks is the difficulty of finding titles for the categories which are neither too detailed nor too slick.

Unless they are closely linked to tested theory, the choice between two or more classificatory schemes remains largely arbitrary. One can, nevertheless, make some judgments on their respective merits *as schemes*. Those which allow for only two classes are inadequate, even if they are sufficiently precise to enable one to classify cities or urban societies unambiguously: by recognizing only two classes, only the crudest breakdown is possible; whereas a scheme which is to be adaptable to a variety of research purposes needs to divide the population in several ways and at a number of points. Similarly, those schemes which are based on vague criteria are inadequate: they may bring out significant elements, but they are of limited value for research if cities or urban societies in a given period were assigned to different classes by different researchers. Finally, a characteristic of the city that is closely related to many others will be more informative than one that is unique. These criteria suggest that there are three complementary bases which can be used profitably for classifying the industrial cities; Reissman's, which describes the society as a whole rather than particular cities; a scheme such as Wirth's, which concentrates on the economic growth and specialization of individual cities, and Moser and Scott's, which underlines the importance of a town's social class structure. The combination of these three schemes, using up to eleven different variables,[40] would give a wide range of possibilities for differing types of analysis. It would enable one to see more clearly how far cities fell naturally into definite categories; and which combinations of these eleven elements were conspicuously absent. It could also have the advantage of relating the features of the city itself to the features of the region or nation in which it is set—a conspicuous omission in most schemes so far.

The situation with regard to pre-industrial cities is less clear. One could, of course, use the same eleven variables, though none of them was designed for pre-industrial cities. As a result, one might overlook a number of important variables in the interest of consistency. The

---

[40] All four of Reissman's, all six of Wirth's, and the social class measure of Moser and Scott. Since this particular combination of variables has not yet been correlated empirically, it might be possible to eliminate one or two of them as superfluous. For analyses within a single country, one could omit Reissman's four variables; though it might be profitable to use them as a basis for classifying *regions* of a country.

classifications designed for pre-industrial cities may escape this criticism; but the most careful of these—by Weber and the Japanese authors—have been drawn up in relation to one area of the world only. Although Sjoberg contends that all pre-industrial cities have a great deal in common, he does not discuss classifications for pre-industrial cities.

(b) Classifications of Social Areas within the City
Several attempts have been made to define and classify distinct social areas within the city. They have been based on the assumption that patterns of land use vary significantly from one area of the city to another; and that one can define areas within which social interaction tends to be warm and intimate (and the residents fairly homogeneous), and between which interaction is formal and reserved. The names of some city streets or areas, indeed, have become synonymous with particular occupational or social groups. At the same time, one would not expect a simple static scheme to remain valid over considerable periods of time: few cities are able to maintain tight control over land use, and the social composition of an area may change radically within a decade. Any scheme which involves mapping the boundaries of the different social areas must contend with constant boundary changes.

Ecologists and geographers have made many studies of land use in particular towns, with the object of devising a general classificatory scheme, and ultimately a general theory of urban land use. This search is perhaps premature: if a complex scheme is needed for categorizing the cities themselves, different types of city may give rise to different classifications of social areas. The social areas found in pre-industrial and in industrial cities, in particular, may have very dissimilar characteristics: in the former, the family workshop or shop may typically be attached to the home, while in the latter the large firm's premises will be clearly separated from its employees' homes. Some types of firm are seriously handicapped if they are unable to obtain sites in the city centre; for others, a suburban or even a rural site may be equally profitable. Planning policies may encourage patterns of land use which conform to a special conception of the public interest, rather than those which would be most profitable for the firms concerned. Governments may reserve prestigeful sites for their own departments, even though from the point of view of economic profitability alone this is unjustifiable. While a general scheme for classifying social areas in all cities does not yet appear feasible, there are plenty of data which could form the basis for a study of land use in cities of particular types.

Before the end of the Second World War, there were three major attempts to devise a model for the industrial city: the concentric zone theory of Burgess, the sector theory of Hoyt, and the multiple nuclei theory of Harris and Ullman.[41] The Burgess theory was an abstraction showing what the city would look like if transport in all directions were good and competition for land was based on economic profitability. There would be five zones: the central business district, with the large stores and the municipal and commercial offices; the transitional zone, where large houses, once inhabited by the rich, have been split into rooms for the poor and the recent immigrants, and are now being taken over by expanding business firms; the zone of artisans' housing, where the general working class population lived; the suburban residential zone; and finally the commuters' zone. These rings around the centre would be modified by siting factors, by the location of industry, by the availability of transport and by cultural values. Burgess' scheme does not apply well to most cities;[42] at the same time it is provocative if it encourages research to discover where and why it is inadequate. Hoyt's theory argued that it was the sector emanating from the city centre, rather than the zone around it, within which homogeneous land use would be found. On this view, one would expect in Britain to find the middle class housing concentrated on one side of the city (often the windward), while the working class housing and the heavy industry tended to be clustered on the other side. Again, site factors, considerations of profitability, transport and cultural values would modify this pattern. In the case of Washington, D.C., for instance, Hoyt recognized that the concentration of the wealthier residents in the North-West sector was partly a response to desirable site features, and partly a self-perpetuating tradition. Finally, Harris and Ullman's theory of multiple nuclei relates best to those cities which have absorbed other towns within their boundaries without changing their patterns of land use: one can then find smaller shopping and manufacturing centres scattered round the central business district, on the sites of formerly independent towns and villages. The envelopment of these smaller centres by the major city may have little impact on land use, especially if the smaller centres are located at important traffic junctions. While each of these theories is illuminating for particular cities, none has proved generally fruitful. Possibly they apply to different types of city; possibly a combina-

[41] Burgess, E. W. 'The Growth of a City'. *Proc. Amer. Sociol. Society*, vol. 18, 1923; U.S. Federal Housing Admin. *The Source and Growth of Residential Neighborhoods in American Cities.* U.S. Govt. Printing Off., 1939; Mayer, R. and Kohn, C. F. *Readings in Urban Geography*. Chicago U.P., 1959; Harris and Ullman, *op. cit.*

[42] See, for example, the critique by Davie, M. R. 'The Pattern of Urban Growth', in Murdock, G. P. (ed.) *Studies in the Science of Society*. Yale U.P., 1937. For more favourable recent assessments, see Schnore, L. F. 'On the Spatial Structure of Cities in the Two Americas', in Hauser and Schnore, *op. cit.*, pp. 349–356; Ericksen, E. G. *Urban Behavior.* Macmillan, 1954, pp. 269–273.

tion of two or three of them would apply better than one alone. Further research to test a more complex theory is needed before this problem can be resolved.

Since the Second World War, there have been few similar attempts to devise a scheme for classifying sections of the city in this way. The 'social area' approach of Shevky, Williams and Bell[43] has been more representative of recent thinking. The search for a global description of each area has been replaced by more careful measurement of the area's significant characteristics. As a result, one's description of an area changes from 'slum' to 'relatively low in house quality, low in income levels, high in incidence of broken homes, above average in overcrowding. . . .' This gives a less evocative but more accurate picture of the area; its weakness is the absence of clear links with any significant theory. During and since the war, various surveys have been concerned to study the catchment areas of different commercial entertainments, clubs, churches, firms, hospitals, and other organizations. Frequently there is only limited overlap between any two catchment areas, but these studies have sometimes offered useful data on the extent to which people are prepared to travel to obtain particular goods or services.[44]

This brief review of the available classifications of social areas has emphasized the difficulty in devising a useful scheme which is not closely linked to a more general theory of urban society. The most penetrating analyses of land use have tended to come from those writers who saw their work in a theoretical context. The urban sociologists have added relatively little: while most of them have conceptualized land use patterns as the results of competition for a scarce resource, they have tended to interpret this too narrowly in terms of economic competition, and then left to the economist the problem of explaining in detail the resultant patterns. Firey's work on Boston is an exception:[45] he showed how an economically powerful group had managed to preserve certain central areas of the city as public parks and historical landmarks, although they occupied sites which made them ideal for commercial development. In general, however, the most consistent and insightful accounts have come from those economists and geographers who linked their analysis of land use to a wider anal-

[43] Shevky, E. and Williams, M. *The Social Areas of Los Angeles*. U. Calif. P., 1949; Shevky, E. and Bell, W. *Social Area Analysis*. Stanford U.P., 1955; Tryon, R. C. *Identification of Social Areas by Cluster Analysis*. U. Calif. P., 1955; Bell W. 'Social Areas: Typology of Urban Neighborhoods', in Sussman, M. B. (ed.) *Community Structure and Analysis*. Crowell, 1959. See also the critical review by Duncan, O. D. of Shevky and Bell, in *Amer. J. Sociol.*, vol. 61, 1955, pp. 84–85.

[44] These include Glass, R. *The Social Background of a Plan*. Routledge and Kegan Paul, 1948; Brennan, T. *Wolverhampton*. Dobson, 1948; and Chapman, D. *The Home and Social Status*. Routledge, 1953.

[45] Firey, W. *Land Use in Central Boston*, Harvard U.P., 1947.

ysis of the reasons for city growth. Haig's analysis,[46] for example, made in the twenties, rested heavily on three economic concepts: the economies and diseconomies of large-scale production; the costs of transportation at various stages in the productive process; and the cost of renting or buying a central site. From these concepts he sought to demonstrate that the reasons for the emergence of the city could also explain the general pattern of land use within it. The location of a particular manufacturing industry was seen as the result of comparing the costs of (i) manufacturing all articles in a single location, and then distributing them to local markets; and (ii) sending the raw materials to the local marketing centres, and manufacturing the products there in smaller factories. This analysis was equally relevant in studying the distribution of facilities for storing goods at various stages in the manufacturing process; or for understanding the concentration of firms' head offices in the centre of the largest city.[47] Residential movements are also susceptible to an explanation in these terms, where the costs of living in a small home close to one's work are compared with the costs of travelling to a more desirable area where one can afford a larger home. In this case, of course, economic considerations are overlaid with value judgments about the social desirability of an area, which may not be closely related to its distance from the city centre.

One looks in vain for a comparable theoretically-based analysis of social areas in terms of sociological variables. A classification of social areas by their density and heterogeneity might be rewarding, but none has yet been attempted. Such a classification would need to take account both of the population which resided in the area and of the population which worked there. Shevky, Williams and Bell categorized social areas as more or less urbanized, in terms of the proportion of the population of the area which is from a minority group, the proportion of married women who work, and the proportion renting rather than owning their homes. While these three measures were derived from theories such as Wirth's, the justification for their selection is by no means satisfactory. Further, the measures may not be transferable to other countries, where values relating to renting, home ownership and employment for married women may be quite

[46] Haig, R. M. 'Toward an Understanding of the Metropolis'. *Quart. J. Econ.*, vol. 40, 1926.

[47] Haig argued that the company executive needs to be able to communicate quickly with bankers, his fellow executives, and the executives of other companies; and that since his time is very valuable, the cost of transporting him into the central city for meetings would far exceed the cost of renting an appropriate, centrally located office for him. In practice, of course, isolation of the firm's head office from its main plants and branches—the price paid for retaining a central location for the head office—makes it more expensive for the executive to tour them.

different.[48] One suspects that, on these measures, the working class in Britain would emerge as necessarily more urbanized than the middle class—whether they were found in the city centre, the fringe, or in a rural area. In order to avoid this objection, it might be more profitable to rely on Lewis's observations,[49] and attempt to define social areas in terms of differential attendance at leisure or work activities which are found mainly if not exclusively in cities. An approach of this type would prevent any simple identification of urbanization with membership of a particular class: the professional football fan and the symphony concert enthusiast would qualify equally as urbanized in so far as they participated equally in activities which were not available to the countryman. Distinct social areas would be said to exist in so far as those who shared common workplaces or patronized common leisure facilities resided in the same area.

V. SUMMARY

Four topics have been covered in this chapter. The first section described Wirth's theory of urbanism as a way of life, and the deductions he made about urban attitudes and urban social organization from his description of the city as a relatively large, dense and heterogeneous permanent settlement. In his view, urbanism was tied, at least in the West, to large-scale economic and political units, to a calculating approach to social relationships, to an elaborate division of labour which obscured simple class differences, and to an emphasis on readily visible symbols and mass-produced goods and services.

The second section noted some of the warnings which Wirth attached to his theory, to point out limits in its applicability and in his intentions. The theory did *not* suppose: that only city residents were urbanized in their ways; that the influence of a town was independent of its proximity to other, larger towns; that size, density and heterogeneity were the only characteristics which cities shared; nor that urbanism always follows the forms which were found in the United States in the twenties and thirties. Finally, Wirth did not attempt, in his article on urbanism, to classify towns or social areas within a town.

The third section examined those criticisms which pointed to weaknesses which Wirth apparently did not recognize. Firstly, some of his generalizations applied only to industrial cities—especially those relating to large-scale organizations and the blurring of class distinctions.

---

[48] In Norway, for example, one of the effects of urbanization between 1900 and 1930 was a *reduction* in the proportion of married women working—partly because urbanization was accompanied by rising standards of living. Since aspirations rose only slowly, many families no longer found it necessary for the wife to go out to work: Norwegian Govt. *Facts about Women in Norway*. Indre Smaalenen, 1960, pp. 31–33.

[49] In Hauser and Schnore, *op. cit.*, pp. 499–501.

Secondly, the rural-urban distinction could not be closely tied to the distinction between primary and secondary groups as Wirth implied. Thirdly, the very variety within the city made many of his propositions over-simplified; indeed, the city may offer wider opportunities for *all* types of relationship than the rural area. Fourthly, Wirth failed to see or explain the persistence of primary groups as an integral part of urban life, and their functions in largely impersonal organizations.

The fourth section has examined some of the classifications schemes for cities and their social areas which might be used to supplement Wirth's scheme for classifying cities. Reissman's four constituents of urbanization in a society form a valuable complement for industrial societies; but no really satisfactory classificatory scheme is available for pre-industrial cities. Schemes for categorizing the social areas of a city have also been examined: while a few types of area can be found in almost all cities, fuller classifications can be made only for pre-industrial and industrial cities separately. There have been a number of promising beginnings, but no fully satisfactory classificatory scheme for social areas is yet available.

The Pre-Industrial City

EMRYS JONES[1] has recently suggested that the popular term 'pre-industrial city' is often used as a residual category, into which all cities that do not fit current Western patterns are placed. It may thus cover mediaeval European cities and all cities of most periods in countries which are not highly industrialized. Doubtless this is in part due to the cultural bias of Western writers; this bias becomes most apparent when the development of cities is assumed to fall into a limited number of logically connected stages, with the folk society representing the one extreme and the highly industrialized society the other. Once this frame of reference is adopted, cities in other types of society need to be represented as intermediate groups along the road or roads. While the early presentations of this argument were very crude, it has recently received more substantial support from Sjoberg;[2] and the work of Reissman[3] might be interpreted as a more sophisticated theory of stages.

Sjoberg argued at length that, as contrasted with industrial cities, pre-industrial cities have many common elements: a technology based on human and animal rather than inanimate power; literacy among a 'leisured elite'; an elite which lives mainly in the cities, though only a minority of the total population lives in cities; an elite which dominates rather autocratically the political, religious and educational structures, and is clearly separated from the lower class and 'outcaste' groups. The upper class maintains an extended family system, and tends to shun economic activity, which is mainly organized through occupational guilds rather than through large firms. Finally, norms tend to be prescriptive rather than permissive.

Reissman took the view that since industrial urbanization implies an advanced technology and a highly developed science, only a limited variety of organizational forms are compatible with it. This places limits on the ways in which a society may become industrialized, since

[1] Jones, E. *Towns and Cities*. O.U.P., 1966, p. 39.
[2] Sjoberg, G. *The Pre-Industrial City*, Free Press, 1960.
[3] Reissman, L. *The Urban Process*. Free Press, 1964, esp. chaps. VII, VIII.

certain necessary conditions must be met. Since these conditions are interrelated and not independent, he argued that the social problems of urbanization tend to be exaggerated unless the meeting of one condition proceeds at a similar pace to the meeting of another. This implies, at the very least, that there needs to be an ideology which supports social change and justifies it as being in the general interest; and a social group which is committed to social change rather than to tradition, and which is able to profit from scientific and technological advances. While this emphasis on common elements in the industrialization process is desirable, it is important to avoid the hasty conclusion, from reading Reissman, that there is little scope for major differences among industrialized societies. It is easy to be impressed by the discovery that all urban-industrial societies are faced with the same problems; and to overlook differences in degree, differences in the cultural setting, and differences in the international situation which will affect the solutions adopted. Similarly, it is easy to be aware of the difference in pace between the industrialization, and especially the urbanization, of eighteenth century England and twentieth century South-East Asia; it requires much more imagination to visualize new forms for the industrial-urban city. Finally, it is easy to conclude that because certain behaviour patterns endure in the West, they must be imperative for successful industrialization; but much harder to judge the range of possible patterns which is compatible with it.

Reissman was careful to avoid the assumption that there is a neat progression of stages through which industrializing countries pass. He recognized the possibility that a country might become relatively advanced in one respect while remaining relatively backward in another; and the possibility that industrial urbanization might stop before it reached an advanced form. Nevertheless, the concept of a limited series of logically related developments is basic to his argument, and has great intellectual appeal. It may be misleading, however, if it becomes ossified by the belief that the four constituent elements of industrial urbanization must always move at about the same pace. Reissman was aware that a number of countries showed a relatively high position on one of his measures, and a relatively low position on another, at the time his data were collected. He was also aware that some of the statistics he was obliged to use for illustrative purposes were untrustworthy. Ar the same time, he said little to account for these deviant cases, and presented no evidence that these societies were being peculiarly disrupted by industrial urbanization. As a result, the consequences of uneven development were by no means obvious; and though different, they may not be more disruptive than those of even development in all four constituents. Even development should not be mistaken for smooth development.

After these prefatory remarks, the chapter has three main sections.

THE PRE-INDUSTRIAL CITY 41

The first brings together some ideas on the rise and development of the city; the focus is on necessary, and if possible on sufficient, conditions for its emergence. The result will therefore be an analysis of common conditions rather than a historical account of causes. The second section discusses some of the principal characteristics of the pre-industrial city: wherever possible, it will be described in relation to Wirth's propositions. Where important aspects of the pre-industrial city are not covered by Wirth, these will be mentioned briefly. The third section examines the conditions under which the pre-industrial city becomes industrialized.

I. THE DEVELOPMENT OF CITIES

This section focuses on the conditions which are necessary before cities can develop fully. Such a task requires a clear conception of a city; an account of the needs of any social group; and the application of this account to the city as a special case. Wherever possible, the necessary conditions will be explicitly related to each other, and a few comments on the more difficult subject of sufficient conditions will be made. Wirth defined the city in terms of relatively large numbers, high densities, and great heterogeneity. To use his definition as the criterion for distinguishing a town or city from a village, one needs to draw up arbitrary dividing-lines: one might, for example, define a city in terms of size, as a grouping with at least 5,000 population, a village as a grouping with less than 5,000. Alternatively, one could make the distinction by introducing additional criteria: defining a city perhaps as a grouping in which most men were engaged in occupations other than agriculture, hunting and fishing, or in which there was an elite whose qualifications included literacy. The more elaborate the definition, the more extensive the list of necessary conditions is likely to be; conversely, the simpler the definition, the shorter the list of necessary conditions and the wider the range of alternative behaviour patterns which could satisfy them.

If Wirth's three key features alone are included in the definition, the dividing-line between village and town must be arbitrary. As the dividing-line in terms of size is increased, the probability of identifying as towns units which are 'really' villages decreases; but the probability of identifying as villages units which are 'really' towns increases. One can only describe settlements as more or less town-like.

Most writers have reformulated the problem, and have asked what conditions and organizational patterns are necessary to support a group in which a significant proportion of the families are not engaged in producing their own food. In effect, this reformulation adds an additional element to Wirth's skeletal definition of the city, by requiring that a significant proportion of the working population be engaged in non-agricultural occupations. Once these principal elements in the definition of the city have been determined, it is not imperative to determine at precisely what point a settlement qualifies as a city. It is, however, important to examine the relationships among the elements in one's definition; whereas most studies have simply considered the conditions under which a non-agricultural population can be supported—assuming that these will also prove to be the conditions for supporting a large, dense, heterogeneous population. The two sets of conditions are by no means identical: the list of conditions given here for supporting a non-agricultural population is by no means a replica of the conditions which Wirth gave as necessary for the stability of a large, dense, heterogeneous population; although of course there is some overlap.

Sociologists have long discussed what needs any social group must meet if it is to survive in its present form; we now turn to this question, as a background for evaluating the necessary conditions for a city to survive. The discussion which follows draws heavily on the ideas of Parsons,[4] but does not purport to be faithful account of his thought. Nor is it intended to support the mis-conception that cities are static groupings, within which little change takes place. Parsons has pointed to four necessary conditions for the survival of a social group or social system in an unchanged form; he implied at times that, taken in conjunction, these may also be a sufficient set of conditions. The four conditions are adaptation, goal attainment, pattern maintenance and integration. Adaptation refers to the need of the group to provide its members with protection against their physical environment, and against hostile groups. This comprises not merely the obvious needs such as food, shelter, control of disease and defence; it includes also the group's attempts to manipulate its environment to its own advantage, by raising production above the minimum, by forming alliances, and by enlisting the services of other groups. Goal attainment refers to the group's efforts to set goals which are desired and attainable, to devise the means for achieving them, to allocate roles and rewards to the members and coordinate their performances, and to justify these goals, means and allocations in terms which the group can accept. Pattern maintenance refers to the defence of group traditions against the desires of new members for change and against the frictions which arise between group and individual interests; it therefore includes the training of new members, control of the effects of the frustrations which arise in their training, the recruitment of new members, and the settlement of disputes among members. Finally, in-

---

[4] Parsons, T. 'General Theory in Sociology', in Merton, R. K., Broom, L. and Cottrell, L. S. *Sociology Today*. Basic Books, 1959; Black, M. (ed.) *The Social Theories of Talcott Parsons*. Prentice-Hall, 1961, esp. chaps. 1, 2.

tegration refers to the development of high morale in the group, of a loyalty to the group which can override self-interest, and of cooperation between members playing different roles. These needs have not been described in full detail; their scope has, however, been indicated.

Within a conceptual framework of this type, one can draw out some of the implications of the attempt to support a significant non-agricultural population. The larger and more heterogeneous this population becomes, the more difficult it becomes to rely on informal arrangements. The denser its settlement, the greater the potential for both conflict and cooperation. Conflict and cooperation should not be seen as alternatives; frequently both are implied by the same joint activity.

Perhaps the most obvious condition for the development of the town or city is the existence of an agricultural surplus. The society needs an economy which is sufficiently productive to release some potentially productive members either seasonally or permanently from collecting or growing food for themselves. Sjoberg argued that in most circumstances this implies a pastoral rather than a hunting basis; and also a technology which includes water control, ploughing and the use of the wheel. Without this technological base, there would normally be insufficient time to achieve more than subsistence, in his view. While an agricultural surplus is clearly a necessary condition, it is not sufficient: the surplus may be consumed or publicly 'squandered' for display purposes; or adherence to a conventional standard of living may mean additional leisure time, which is then used for activities which provide, say, increased relaxation rather than increased output. Many developing countries have high rates of 'concealed unemployment'.[5] The surplus may be used to support craftsmen, traders, inventors or an elite class which does not contribute directly to production. While this concept is clear in the abstract, in practice it may be very difficult to measure except where a significant proportion of the population is not engaged in agriculture.

If part of the surplus is used to support non-agricultural urban groups, a number of conditions are necessary for social stability. Firstly, the agricultural producers must be induced to hand over their surplus. This might be achieved through military and political domination, by either killing or subduing and taxing the producers. The agricultural producers might also be induced to continue producing and handing over their surplus in exchange for other commodities or

---

[5] I.e., a situation in which everyone who desires work nominally has a full-time job, but actually works at it only on a part-time basis. The peasant family which uses three members to tend three animals, where one could perform the work with existing tools, is an example.

services which they find useful.[6] Frequently the residents of small towns in an agricultural area perform direct services for the agricultural producer. Clearly the scope for such services increases as the producers' equipment and aspirations become more complex, and as the surplus itself increases. The services provided are not necessarily technical, of course: some are connected with the organization of a market where the surpluses can be exchanged; others with the provision of entertainment and relaxation for the producers; still others with the organization of religious, political and educational groups to which the producers may belong. The services may, indeed, combine elements of all these types: there may, for example, be religious leaders who receive some of the surplus and who give in exchange an elaborate and rationalized production routine. Such a routine might include religiously based decisions on when to plant or harvest, how to ensure good crops, how to intercede with the gods in controlling the weather, and how to justify crop failures.

Secondly, the existence of a surplus which is used to support urban life necessitates methods of transporting it safely to the town. While this has been emphasized by some writers, others have made it clear that efficient transportation is necessary only where considerable distances have to be covered. This condition is significant, then, in that poor transport restricts the distances over which goods can be transported, and correspondingly limits the hinterland and potential growth of a town. Poor transport prevents the town from using the surplus from any but the most local production; this limits severely both the non-agricultural population which can be supported, and the extent to which the town can develop an advanced division of labour.

Thirdly, the surplus needs to be stored and protected, which requires additional facilities and labour; and if an exchange system more complex than the crudest barter is involved, measurement becomes necessary. A system of weights and measures is needed; and if supervision of the granaries is delegated to an official, he is likely to be required to keep accounts. Where the surplus is handed over as a form of taxation on land, elements of geometry are needed for the calculation of land areas. Fourthly and finally, facilities for distributing the surplus to the town population are needed. This may be done through the direct exchange of the townsman's service for the countryman's produce. Exchanges of this kind can be satisfactory as long as either the service or the product can be divided into very small amounts; but are much less feasible with large, indivisible goods or services. It quickly becomes unfeasible when more than two parties

---

[6] For an analysis of exchange and exploitation in more detail, see Keyfitz, N. 'Political-Economic Aspects of Urbanization in South and South-East Asia', in Hauser, P. M. and Schnore, L. F. *The Study of Urbanization*. Wiley, 1965.

are involved in a transaction, or when the townsman no longer needs more produce; and the development of indirect services is dependent on a common medium of exchange, which need not of course be money itself. The institutional arrangements which accompany the distribution and exchange of the agricultural surplus thus have far-reaching influences; and conversely the necessary conditions for using an agricultural surplus to support a town population are complex.

In addition to food supplies, the town also needs access to a relatively large and steady water supply, and the basic raw materials for assuring shelter and defence. While these needs are not of course unique to the town, its greater size and density increases the amount and possibly the variety of materials required.

The city population's other areas of need—goal attainment, pattern maintenance, and integration—have not been treated so thoroughly; but this should not lead one to underestimate their importance. With regard to goal attainment, the city dwellers need a system for distributing power and rewards, and for justifying the inequalities which result, whether intentional or not. Since more than one occupation is represented in the town, and since power can no longer be equated with the extent and productivity of one's land, another basis, such as control over military manpower or the value placed on the services rendered, needs to be found. Since roles in the town tend to be varied and not readily comparable, mechanisms are needed for controlling competition in the provision of services and for fixing the prices of these services.

In the area of pattern maintenance, the recruitment and training of persons to continue performing agricultural services, the maintenance of order and routines, and the settlement of disputes between and within occupational groups are important needs. Needs of these two types can frequently be met through the establishment of a market, in which goods and services are sold competitively, and prices are fixed at the end of a bargaining process. While this appears to be a simple and rational solution, closer inspection reveals that it depends upon a series of conditions, in addition to the usual economic conditions.[7] It depends, for example, upon a general willingness to follow the 'rules of the game', even in situations where deviance might be profitable. It depends upon the participants' willingness and ability to spend considerable time in judging the prices of products and services carefully, so that one may detect small differences between them. It de-

---

[7] Traditional economic theory was geared to combinations of three assumptions, which could relate to either the buyer's or the seller's position: (i) that there were so many buyers (or sellers) that no-one was able to influence the market price by withholding his patronage (or produce); (ii) that there was only one buyer (or seller); (iii) that there was more than one buyer (or seller), but few enough so that the larger buyers (or sellers) could influence the market price.

pends upon free movement, and upon the restraint of the richer participants, in not using their political and religious power as bargaining tools; and on a willingness to deal with newcomers if they offer a better bargain. It depends on the preservation of a clear distinction between economic and other relationships, so that kinship, personal liking or animosity, class and religious differences do not influence the course of bargaining or the preference for one trader rather than another. The solution of the economic problems of the city thus requires a political framework, and considerable agreement on goals. Sjoberg, indeed, argues that the political framework usually precedes the spread of trade; and political power rather than commerce is the key to the rise and spread of urban centres.[8]

Lastly, the need for integration implies a loyalty to the city and its constituent groups, to its religious observances and political institutions. In the middle ages, the formal expression of this loyalty often took the form of an oath of allegiance, usually in the towns to an occupational guild or to a mutual protection society (*conjuratio*).[9] The bonds within these guilds and mutual protection societies frequently replaced the bonds between a lord and his dependents in the rural areas. The craftsmen and later the merchants of the towns gradually developed different ideologies from the agricultural producers, and came to think of themselves primarily as townsfolk. These differences perhaps became clearest in traditional Indian society, where the craftsmen and merchants were treated as members of different castes. Loyalty to the town and to one's occupational group also implies efforts to maintain them, not merely against open attack but also against economic decline. One therefore found that the groups which obtained power tended to increase the number of persons who were directly dependent upon them. In this way, occupational groups gradually arose which provided goods and services for their fellow townsmen, rather than for the agricultural producers; this tended to happen in so far as the agricultural surplus could be increased beyond the minimum needs of the current town population.

The argument so far has had two main purposes: to elaborate some of the necessary conditions for the maintenance of towns in an agrarian society, and to suggest that these conditions become rapidly more complex as the town grows in size and heterogeneity. In the earlier and simpler towns, many of these needs appear to have been met in the same manner: power was concentrated in the hands of a few leaders, who exercised it in religion, politics and most other areas. Where they were a hereditary group, problems of succession could often be decided in a relatively routine manner. Since they were the main sources of re-

---

[8] Sjoberg, *op. cit.*, pp. 67–77.
[9] Weber, M. *The City*. (Eng. trans.) Free Press, 1960, pp. 94, 106–107, 110.

ligious knowledge, and of the group's beliefs; and since they were the founts of its rituals and routines, they had the backing of traditional authority, and were the patrons of those officials who had such skills as literacy and numeracy.

Finally, we must consider the possibility that towns set up as fortresses and administrative centres under a colonial regime are quite different from towns which represent the outgrowth of a local agricultural surplus. These towns may have only limited trading ties with their surroundings; and may be dependent on superior techniques or on rapid and reliable communication with the founding city for their survival. Relations with the surrounding area are likely to be tense but ambivalent. On the one hand, the garrison town is likely to be supported from local produce and local taxation, and manned by 'foreigners'; on the other, its existence may provide alternative sources of employment and gain for local residents. There may be few obvious constraints on the methods which the garrison uses to control the countrymen; at the same time, it is necessary that they be restrained enough to retain at least a modicum of cooperation from their subjects.

None of these conditions which appear to be necessary for the maintenance of a town suggest themselves readily as sufficient conditions; if sufficient conditions exist, it seems likely that they will consist of sets of many necessary conditions. Since size and density simultaneously increase the opportunities for both cooperation and friction, the necessary conditions for stability are likely to be more numerous and perhaps less closely interdependent for the town than for the small village. At the same time, the greater heterogeneity of the town decreases the likelihood that integration can be maintained on the basis of shared values and similar characteristics among the population; and increases the extent to which integration depends on the recognition of the gains which cooperation and a complex division of labour bring. The increasing complexity also makes it more difficult for the voice of traditional authority to speak clearly to all groups; and thereby increases the need for voluntary collaboration in the maintenance of the city's social order.

II. THE PRE-INDUSTRIAL CITY: A BRIEF PROFILE

The second section of this chapter outlines some of the main findings about the social organization of the pre-industrial city. Wirth's propositions form the framework around which the material is organized; but if these prove inadequate, they will be supplemented. Sjoberg's work forms the most complete single source on the pre-industrial city; though since his objective was to present common elements, he makes relatively little effort to develop a classification of differing types of pre-industrial city. This task was attempted by Weber for European and classical cities,[10] but has rarely been systematically attempted elsewhere. As a result, the emphasis here will be on elements which pre-industrial cities share; and little attention will be paid to divergences among them.

Proposition 1. *Relatively weak bonds among co-residents; formal social control; and/or physical separation of diverse sub-groups.* It was not very clear from Sjoberg's account whether bonds among lower-class and 'outcaste' co-residents are strong or weak as a rule; though one's impression was that strength is found only when co-residence is supplemented by work in the same occupation, and by a low level of residential mobility. Where this is the case, loyalty to a common occupational guild is frequently strong; but where there is no occupational bond, co-residents may indeed have little in common. Among the elite or upper class, on the other hand, strong bonds based on kinship and common interest are readily noticeable: they occupy exclusively the central residential area, interact regularly and normally marry only within their own social class. Social control over other groups is maintained through a variety of informers, officials, troops and sometimes through magistrates. These functionaries are the agents of social control, and there is little direct contact between the elite and the lower class. The authority on which their positions rest is justified in terms of written tradition and moral absolutes, rather than in terms of political expediency, rationality or outstanding personal qualities. To challenge the authority of the elite would therefore threaten the foundations of the social order; even if the struggle against poverty had left the lower class with the time and strength to make such a challenge, prevailing beliefs would have stressed its immorality rather than its inexpediency. This challenge might have been avoided on the grounds that it would be ineffective; more assuredly it would have been discouraged as shocking and reprehensible.

The physical separation of diverse sub-groups is evident. Not only is there a clear distinction between the central plaza with its market, religious buildings, and open space for ceremonials, and the remainder of the town; in addition, the social classes are segregated residentially. The elite often takes steps to keep the 'outcastes' away from the city centre altogether; and in any case many of the lower class choose to settle near the edge of the town, where they have better access to cultivable land to supplement their income. Where the town contains diverse ethnic and religious groups, each tends to have its own quarter, where group members both reside and work: these quarters may even be surrounded by walls to accentuate the differences and further reduce contact and possible friction. Finally, one finds segrega-

---

[10] *Ibid.*, pp. 197–223.

tion according to occupation, in cases where there are no major ethnic or religious differences within the population: the producers of a particular good or service tend to congregate within, and to monopolize, a particular quarter or section of a street.

Proposition 2. *Difficulty of knowing all others personally: impersonality, superficiality, etc.; relationships treated as means to ends.* The first part of this proposition holds true almost by definition. Relationships among the elite are again far from impersonal and transitory; but casual acquaintanceships are probably common among the general population. Before one accepts this characterization of commercial relationships in the pre-industrial city, however, two warnings are necessary. Firstly, trader and customer may both gain considerable benefits from developing a personal relationship: the trader is better able to rely on his customer's continuing patronage, and has a chance to obtain that of the customer's relatives, which is important in a market system where aggressive sales promotion is largely unknown.[11] It offers the customer protection against the unscrupulous trader if his very limited purchasing power is backed by that of his kin, and if personal obligations are imposed on the commercial relationship. Further, since time is not a scarce commodity, conversation and trading is a pleasurable activity in itself. Secondly, the commercial relationship is usually surrounded by highly institutionalized forms of behaviour: courtesies, derogation and exaggeration of the product's virtues, and haggling over price. Indeed, Sjoberg argues that the trader is careful *not* to sell all his goods early, because it is unthinkable that he would be able to use the time saved to increase his stock and his turnover. While these modifications are necessary, commercial relationships do contain an element of exploitation, in that the financial gain of one party represents a loss for the other, and each can take pride in outwitting the other. In a situation where standardization of weights and measures and commodities is largely absent, there are few reliable methods for enforcing honesty among either traders or purchasers. On the other hand, the rewards are frequently small, since most transactions involve two very poor persons.

Proposition 3. *Division of labour associated with treating relationships as means to an end; dominance of the large firm; codes of ethics for occupational and other specialized groups.* While there may be a very clear division of labour, the division in the pre-industrial city is by product and not by process. There may be many similar commodities on the market, each made by a separate guild, which resides and

---

[11] This is not to imply that the trader is not interested in selling his goods and services eagerly and at a handsome profit. It does mean that there is limited opportunity for *increasing the volume of sales* through advertising and sales campaigns, because there is little opportunity for obtaining, storing and distributing goods in bulk.

D

works in a different part of the town; but apart from the division between master and apprentice, the manufacture of a particular product is unlikely to be subdivided into many parts. The production line is unknown: the same man may buy the raw materials, fashion the object and then sell it. The guilds into which many occupations are organized often set up rigid standards with regard to entry, quality of product, selling practices, and minimum prices, and demand loyalty from their members; there are therefore specialized codes of ethics.

Large-scale production is conspicuously absent, however, from this pattern of urban life; many of the necessary conditions for it are apt to be unmet. There is, for example, little opportunity for bulk purchasing, storage or selling. Indeed, Sjoberg argues that since the economy is organized for trade in small quantities, the purchase of materials in bulk would be disproportionately expensive. Similarly, there are few opportunities for obtaining credit. While the elite buys from the merchants, it regards them with disdain because their work involves association with the lower classes and some manual labour; and therefore the elite neither invests in their enterprises nor trades directly itself. Further, in mediaeval Christianity and Hinduism, at least, loans at interest were discouraged on religious grounds. Where loans are available, the trader rarely has much collateral to offer. As a result, the trader who needs credit must rely on ploughing back the profits from his business, and on help from his kin. Neither of these sources can usually offer more than token amounts of capital. When these difficulties are allied with a simple level of technology, which makes mass production less feasible, a slow pace of technical advance, and the general poverty of the city's lower class and outcastes who represent the only potential mass market, the problems of investment and capital formation are made even more acute. Traders can become rich only by dealing in luxury goods for the elite; and since the elite are few in number and expect individually tailored goods, trade with the elite does not form a foundation on which large-scale industry can be built.

Proposition 4. *The division of labour grows with the market; extreme specialization, interdependence, unstable equilibrium.* Available data from the pre-industrial cities can neither support nor deny this proposition clearly, because it is couched in terms of changes over time; whereas most of our information is on single points in time. Certainly there is specialization and interdependence to some extent; but the markets grow only slowly. Little attempt is made to expand the volume of trade, especially if this involves hard manual work; and the merchant's tempting appeals are addressed to an individual rather than to a mass audience. Interdependence is certainly most marked among the elite families, who may have many retainers.

Proposition 5. *Impossible to assemble all residents together; indirect communication by mass media and representatives of special interests.* While there is very likely to be a central square in which large-scale ceremonies are held, this is frequently too small to contain all the citizens. The result tends to be erratic communication, rather than the kinds which Wirth expected. There are normally no mass media in the Western sense; though there may be officials charged with making public pronouncements at the tops of their voices. Communication within the elite, within a guild or within an extended family may be efficient; but between groups it tends to be much slower. Communication between towns depends on travelling merchants, story-tellers and other itinerants: again this makes news delivery unpredictable and limits the radius of influence which a town has.

Proposition 6. *Differentiation and specialization increase as density grows.* Again, we have no reliable evidence from studies over time.

Proposition 7. *Physical contact is close, social contact is superficial; people are categorized by readily perceptible symbols.* Social and physical contacts between members of different groups are lessened by residential segregation, but not altogether removed. Indeed, in an environment where the streets are frequently those strips of land where no-one has yet built, and often not more than two or three feet wide, many physical contacts are unavoidable. Similarly, only the elite can afford houses with grounds and privacy. The result of close physical contact and limited social contact is not anonymity, however. Members of the elite are readily distinguishable by their dress, speech and manners. Their avoidance of manual work and use of scents will accentuate these differences. Variations within the lower class or among outcastes may be less striking than those between the elite and the rest of the population. In some instances, nevertheless, each guild may wear its own symbolic garment; and in traditional Peking the dress and instruments of the peddler represented symbolically the article which he was selling. As was noted under Proposition 2, the impression of urban social relationships as superficial needs to be modified: personal ties may be important in trade as in other activities, and social relationships may follow carefully prescribed forms.

Proposition 8. *Land use is the result of economic competition; residential desirability is a complex concept; residential segregation is usual.* While land use in the pre-industrial city is symbolic as well as utilitarian, the symbols are based on political and religious rather than on economic criteria. The city centre usually contains the principal market, but it would be misleading to refer to this area as the central business district; its main buildings are more likely to be religious, political and ceremonial. The main reason for the central location of the market is probably the proximity of that area to the homes and work-places of the elite and their officials, who have the greatest purchasing power; it is less likely to be due to the convergence of all main transport routes on the centre. It would also be misleading to imagine the majority of the residents taking regular shopping trips to the central market, or to suppose that economic considerations outweigh political and religious interests when the two clash.

Residential desirability is directly related to distance from the centre: the most valuable sites, those chosen by the elite, are in the middle of town, which is the best protected area, and the best situated to take advantage of urban amenities and to maintain close contact with other elite families. As long as transport remains poor, and constant communication is feasible only over short distances, the elite tends to reside in the central areas; and increasing poverty is associated with increasing distance from the centre. The surrounding areas are characterized by very high densities, and narrow twisting streets which are poorly drained and rarely kept clean. The lower class and outcastes typically live in different parts of the town, according to their ethnic, religious and occupational affiliations; each 'ghetto' may be surrounded by its own wall. There is, however, very little segregation of residences into some parts of the town, and of workplaces into others. For most lower-class town-dwellers, the home and the shop or workshop are part of the same structure; they may, indeed, be the same room. The segregation of home and work may be more pronounced for the outcastes, though Sjoberg is not clear on this point. Since many of them are labourers or itinerant workers, they obviously do not work on their own premises; but one is not certain whether they have an appreciable journey to work, in the Western sense.

Proposition 9. *Competition and mutual exploitation rather than cooperation at work; life at a fast pace; orderly and meaningful routines needed.* Wirth's picture of modern production as competitive and exploitative does not necessarily apply to production in the pre-industrial city. Firstly, there is much less stress there on individual initiative and personal gain; and the links between increased productivity and higher earnings are much less readily apparent. Loyalty rather than competition is stressed by the guilds, and cooperative efforts are sometimes made to help a member in distress. The guilds' function may be to eliminate competition as far as possible, by using kinship as the basis for admission and setting conservative goals which restrict ambition. Ambition and personal gain may be further restricted if there is a dominant emphasis on sharing one's profits with kin and other guild members. Secondly, business dealings are surrounded by rituals which are time-consuming and which are more effective in maintaining a stable market than in increasing turnover. At the same time, one should not conclude that exploitation is unknown: the long-established guild members may exploit the newer members, and especially the apprentices, in many ways; and the latter may look forward to exploiting their own apprentices in the future. Similarly, the buyer may be regarded as a legitimate subject for exploitation: the guild's code of ethics may have little to say on relationships with buyers, except to ban price-cutting.

Proposition 10. *Simple class distinctions break down as the division of labour becomes complex.* While the division of labour in pre-industrial cities may sometimes be complex, distinctions between the three broad classes remain very clear, according to Sjoberg. In addition to the readily perceptible differences already outlined, the elite alone are literate; and they live separate from the masses and interact little with them. The distinction between the lower class and the outcastes is less striking, but Sjoberg argued that it is firmly maintained. He stressed that the lower class is able to be proud of the 'respectability' of its occupations, by contrast with the 'degrading' work which is done by the outcastes. Within each class, there may be sub-divisions which are more difficult to distinguish readily.

Proposition 11. *Group loyalties conflict; geographical and social mobility is pronounced; sophistication results.* The conflicts of group loyalties are much less marked than in industrial societies, because organizational membership is much more likely to be ascriptive. There are very considerable overlaps between kinship, occupational, religious and educational groups, even among those classes where the extended family system is relatively weak. Among the elite, the coincidence of these groups is most marked. Social mobility is rare, though by no means unknown. Sjoberg's work gives the impression that downward mobility may be more common than upward mobility. Even if fertility is lower among the elite, its death-rates are so much lower that its more rapid growth would threaten the maintenance of its relatively high living standards. Among the lower classes, by contrast, death rates are usually high, and may be extremely high during an epidemic. There may be rules of inheritance under which the younger children of an upper class family, and especially the younger sons of younger sons, are found jobs with only semi-elite status. Similarly some of the sons of officials may have difficulty in finding jobs comparable to their fathers'; and their own children may eventually enter lower-class occupations. Upward mobility is unusual, because it depends on radical changes in style of life, and not merely upon the accumulation of wealth through trade or power through military success. There is little evidence to suggest that conflicting group loyalties are a major source of sophistication in Wirth's sense; from the more varied social contacts which their life involves, one might expect to find greater sophistication among the lower classes and the outcastes than among the elite.

Proposition 12. *Levelling influence of mass production; pecuniary nexus.* The literate elite, which has the skill and position to coordinate a more elaborate division of labour, is contemptuous of manufacture and trade. Inability to use inanimate sources of power, the poverty of consumers, the problems of accumulating capital, and the low level of technology, all make mass production difficult. As a result, the standardization of goods and services which it involves does not develop; indeed, the weights and measures used may themselves be highly variable. The development of the pecuniary nexus is also less apparent, because many services are still performed within the kinship group, and for this reason the possibility of valuing them in monetary terms is much less likely to arise. If valuation should be necessary, this is more likely to be made on the basis of religious and political criteria.

This account of the pre-industrial city has drawn heavily on Sjoberg's comparative study, and on Bascomb's studies of Yoruba towns.[12] The picture presented is by no means remote from British experience, however. Although Britain is the most urbanized nation in the world, one can still find traces of the pre-industrial city. Perhaps the most striking is Margaret Stacey's portrait of the traditionalists in Banbury.[13] Obviously there are differences, in that Banbury is not afflicted with chronic poverty, and has regular access to the mass media. Yet the traditionalists could be fitted fairly precisely into this account without major modifications.

It can readily be seen from this account that some of Wirth's propositions apply poorly to the pre-industrial city, while others need modification. There are two obvious reasons for this: firstly, Wirth had in mind the very large, though not necessarily the industrial, city. He commented that some contemporary non-Western cities should be viewed as large villages, for the purposes of his analysis. Consequently his propositions ought to be more applicable to the very large pre-industrial cities than to the relatively small ones which are in the majority. Regrettably, there are insufficient studies of the largest pre-industrial cities to establish whether Wirth's propositions would apply better to them. Secondly, there were few sound studies of pre-industrial cities at the time Wirth was writing; consequently his theory could not describe the process of urbinization very fully. It is evident from the material on pre-industrial cities that the combination of large-scale commercial enterprise and the elaborate division of labour is only one possible result of increasing size; it is also possible for city growth to result in a proliferation of small firms, coupled with an elaborate governmental organization in which political, religious, economic and other sources of power are concentrated in the hands of an elite. Similarly, Wirth's arguments rested upon the tacit assumption that transport and communication were efficient, whence it followed that those who worked together need not share other aspects of their

[12] Bascomb, W. R. 'The Urban African and his World'. *Cahiers d'Etudes Africaines*, vol. 4, 1963.
[13] Stacey M. *Tradition and Change*. O.U.P., 1960.

THE PRE-INDUSTRIAL CITY 55

lives unless they chose to do so; and that the elite could maintain constant communication without living together. One further tacit assumption needs to be eliminated: the assumption that the culture's values need not be introduced into the analysis as causal variables, since these can be deduced from size, density and heterogeneity. This assumption meant that Wirth overlooked the possibility that traditionally-oriented cities could develop, in which social control was maintained by tradition and dogma, rather than by formal means of inspection, by clocks and by regular precise time-tables; cities in which economic and scientific progress was valued only in so far as it contributed to the elite and the stability of the society. Once these unstated assumptions are removed, Wirth's propositions offer a very useful basis for understanding the pre-industrial city. In conclusion, it should perhaps again be stressed that this account has deliberately concentrated on the elements which most pre-industrial cities have in common, and has tended to overlook differences among them which a fuller analysis would be obliged to include.

III. NECESSARY CONDITIONS FOR INDUSTRIALIZATION

The final section of this chapter is designed to bridge the gap between the pre-industrial city and the industrial city, by commenting on some of the conditions which appear to be necessary for the development of industrialization. The preceding section presented a somewhat static view of the pre-industrial city, and pointed to some of the obstacles to industrial development and large-scale capital formation. While the pre-industrial city proved stable over centuries, it would be misleading to leave the impression that its social patterns were unshakeable. Historically, the process of industrialization has tended to occur in two ways. In Western Europe and North America, it developed after a long period of gradual economic expansion and increasing scientific knowledge, and can perhaps best be described as an indigenous development. In many other parts of the world, industrialization has been largely an imported growth, which has tended to spread rapidly wherever social conditions were favourable. This rapid spread has shown the inadequacy of attempts to deduce the conditions for industrial urbanization from a study of Western experience alone; and leaves open the possibility that the conditions for the *origin* of industrialization may differ from the conditions for its *spread*.

Full industrialization has several related components: the harnessing of inanimate sources of power to work machinery; the introduction of large-scale production; the division of labour by process; and production according to a precisely determined schedule. It is valued both as a goal in itself and as a means of achieving other goals such as higher living standards, health and national power. The conditions which are necessary for industrialization have received a great deal of attention over the last twenty years; this section will indicate some of the factors which have been mentioned most frequently. The emphasis here will be on the institutional arrangements which need to develop alongside industrialization; there is no suggestion that these need to be fully present before industrialization can begin, but it is argued that they are necessary accompaniments if a stable system is to result. They may best be seen as factors whose absence, singly or jointly, raises serious obstacles to the full development of industrial cities in the Western sense; though at the beginning of industrialization they may be only weakly developed themselves.

The first three factors relate to the organization of large-scale production itself. Mass production by the use of machinery and inanimate power requires capital, readily available mineral resources, and considerable technical knowledge. There is no intrinsic reason why the most useful minerals should be found, or the major scientific discoveries should be made, in the richer farmlands, which are the most likely locations for early urban development. As some of the other conditions for industrialization are met, of course, the utilization of natural resources becomes more efficient because the acquisition of technical knowledge, the accumulation of capital, and the exploitation of existing resources encounter fewer obstacles. These obstacles to industrial development are interrelated and may reinforce each other; but there is little to suggest that they are related to each other in a simple manner. As a result, one cannot represent industrialization as a series of neat, logical stages, each of which represents the removal of a single obstacle.[14] It is in some ways more helpful to regard the process as akin to the expansion of a many-sided hollow figure, whose sides are flexible and made of different materials. The figure can be inflated by exerting pressure on any one or more sides; some will recede readily, while others may be more resistant. Pressure on one will have repercussions on the others; the final shape of the figure will depend to some extent on where the pressure has been exerted most forcefully and most consistently. This analogy may also be helpful in reminding one that movement towards industrialization in one respect may lead to movement away from industrialization in another. The analogy represents only a first step, however: no shape for the figure has been specified, and no attempt has been made to examine the limitations on final shape imposed by the interconnections between the many sides of the figure. Only as these additional limits are incorporated can one

[14] Indeed, several writers have explicitly warned against acceptance of the idea that industrialization imposes such constraints on organization that the most industrialized cities become increasingly similar. See Halmos, P. (ed.) *The Development of Industrial Societies*. Sociological Review Monograph No. 8. Keele, 1964.

expect a fuller understanding of industrialization, and the processes by which it may be attained.

The second factor relating to the organization of large-scale production is the credit market. The delays between the purchase of the machinery and the sales of its products are considerable, credit is essential wherever a manufacturer has insufficient capital of his own. The existence of organizations which can collect and distribute capital in turn depends on a number of prerequisites. Saving is possible only when the agricultural economy is able to produce a considerable surplus; the investment market must be sufficiently well-known, reliable and secure to induce men to invest rather than to hoard their savings; and investment in industry must be attractive relative to other uses for savings. These include not only investment in agriculture and war adventures; but also such expenditures as are incurred in the erection of religious monuments, the support of the arts or the acquisition of luxury goods.

The third factor relating to the organization of production is the existence of an efficient transportation and communication network. This is necessary to ensure that supplies of raw materials in bulk can be obtained at short notice, that technical knowledge spreads rapidly, and that technical training is available to meet manpower needs. Education is included here as part of the network of communication; for large-scale production, this involves both technical education for particular occupations and the general education which is necessary for performance of even simple factory jobs. Efficient transportation gives access to a much larger market, and is especially important in marketing perishable commodities. It also permits the more intensive concentration of population in large cities, which in turn makes the exploitation of a large market and the economies of scale more feasible.

Fourthly, a moderate degree of urbanization is probably a necessary condition for advanced industrialization. Breese has argued that either industrialization or urbanization can begin without the other,[15] but that continued development is probably contingent on the occurrence of both simultaneously. Lampard, however, reached the conclusion that 'large cities do not require a developed economy; but a developed economy requires a hierarchy of cities'.[16] While there have been large pre-industrial cities, some perhaps with populations over one million, these have, so far as one can judge, been 'primate cities': administrative and commercial capitals in largely rural societies, which had no other cities of remotely comparable size. Obviously, then, one may have urban growth without industrialization; but there

[15] Breese, G. E. *Urbanization in Newly Developing Countries*. Prentice-Hall, 1966, pp. 51 *et. seq.*
[16] Lampard, E. E. 'The History of Cities in Economically Advanced Areas'. *Econ. Develop. & Cultural Change*, vol. 3, 1955, p. 82.

is little evidence that industrialization can occur without both urban growth and urbanization.[17] Historically, industrialization has tended to occur only after urbanization was under way; more recently, it has been expanded and introduced deliberately in non-Western countries as a method of solving the problems of rapid urbanization. As the belief in their interconnectedness grows, of course, it becomes increasingly difficult to test whether there is any necessary relationship between the two.

The remaining factors are mainly concerned with the institutional setting and the values which are necessary for the full development of industrialization. Moore[18] examined some of the forms which economic institutions take, and concluded that certain forms are especially conducive to industrialization and may form necessary, though not sufficient, conditions. As one condition, he mentioned a set of property norms which effectively separate most workers from private ownership of the means of production. This is not a judgment on the relative merits of efficacy of public and capitalist ownership of large firms. Moore's point was, rather, that industrialization requires the development of large firms which can own the large amounts of capital and can organize the large quantities of labour that mass production implies. Whether these firms in turn are privately, cooperatively or publicly owned does not, in his view, affect this argument. The system which he found most opposed to industrial development was a system of peasant proprietorship, wherein each family has rights to the produce from a small parcel of land, and owns or rents its own simple tools. Property rights of this latter type make the accumulation of industrial capital very unlikely.

Industrialization also requires, according to Moore, a system of labour norms which encourage freedom and experimentation. This implies willingness to try new ideas and take risks, to learn new skills and to move from place to place in search of higher economic rewards, to risk specializing in a single skill or product, to work in a large and relatively impersonal organization which has a variety of formal routines. One could regard these norms, both as the types of behaviour which an industrial society encourages, and as the type of personality which is most likely to lead to success in that society. Again, many pre-industrial societies lack most of these norms: attachment to one's traditional plot of land is the rule; any agricultural surplus is con-

---

[17] Some writers have distinguished carefully between these two terms, to allow for the existence of primate cities. They have used 'urban growth' to refer to the growth of particular towns, and 'urbanization' to refer to the process by which most members of the society come to live in towns. In the primate city case, urban growth has taken place; but if 80% of the population still lives in rural areas, the society is not urbanized.

[18] Moore, W. E. *The Impact of Industry*. Prentice-Hall, 1965, chaps. 3, 4.

sumed in leisure pursuits which are not economically profitable; and systematic commercial enterprise (as distinct from military plunder or shrewd bargaining) is regarded as beneath the dignity of those who have sufficient capital to engage in it. Similarly, in many pre-industrial societies, the ties of kinship and the obligations to guilds and overlords are so localized that labour mobility is effectively restricted.

Moore also argues that industrialization requires a system of exchange which is based on the impersonal marketing of goods and services, and on the determination of profits, salaries and wages in monetary terms rather than in kind. Again, his emphasis is on the removal of restrictions to free commercial enterprise which are represented by traditional local ties. This emphasis could be misleading in several ways; and it is important not to identify the necessary conditions with the conditions which happen to hold in any one industrial country. Firstly, one should be careful not to draw hasty conclusions about the relative merits of capitalism, cooperative ownership and socialism, by assuming that one form of ownership of industry is peculiarly well adapted to industrial development; nor should one make general judgments on the relative merits of oligopoly and perfect competition in this respect. Secondly, it should not be supposed that the endless removal of restrictions will lead to ever-increasing industrialization: Moore also made it clear that a certain amount of social and political stability is necessary. Since there are considerable delays before the full benefits of an investment can be enjoyed, the investor needs protection against expropriation and radical political change. Where the chances of protection appear to be slender, only very high rewards will attract the investor.[19] Finally, the maintenance of any particular kind of economic organization may require government support and intervention. Perfect competition, for example, requires the backing of laws or regulations which govern the means by which rival firms may compete; otherwise whenever there are economies of large-scale production, the unscrupulous competitor can gain a dominant market position and then use it to suppress potential competitors.

Turning now to values, it is evident that an industrial society differs considerably from a pre-industrial society; and it is arguable that certain values are necessary for full industrialization. Reissman,[20] for example, mentions a nationalistic ideology which favours the replacement of colonial and tribal by national administration, and which justifies this in terms of benefits to all groups. While nationalism (and

---

[19] Again, mass production normally implies the gathering of resources which are widely scattered; reasonable political stability is one of the factors which helps to ensure a steady source of supplies.

[20] Reissman, L. *The Urban Process*. Free Press, 1964, pp. 180–188.

more often, anti-colonialism) has frequently accompanied the spread of industrialization from the West, it should not be supposed that this is the only possible ideology. Protestantism may have had similar consequences in the West, albeit unintentionally.[21] The critical features of the ideology seem to be an appeal for radical reform in the political and possibly also in the religious institutions, and an appeal to the common interests of groupings which would be wide enough to form a mass market if they were socially and politically unified. Clearly the Reformation was not primarily a political movement, though it did strike at the waning political power of the Roman Catholic Church; and of course it was not primarily a movement to increase markets to the point where mass production became feasible. Nor could it be classified as a forward-looking reform movement, since much of its objective was a return to traditional practices and virtues. Nevertheless, it attempted to unify religiously a considerable segment of the world's population; to some extent it succeeded in doing this and in providing a legitimation for both missionary work and colonialism. Protestantism also motivated many, according to Weber, to enter business and to make money diligently; since it discouraged them from spending this money on themselves or on others, most of the revenue was ploughed back into their businesses and overcame some of the traditional problems of capital formation. In time, the business activity became dissociated from Protestantism, and acquired a momentum of its own.

While Protestantism in the West and anti-colonialism in developing countries have historically been associated with industrialization, neither is in itself a sufficient condition. Anti-colonialism, for example, is not intrinsically and inevitably linked to industrialization, but could favour many other forms of social change. It is highly relevant only when it becomes associated with or develops at the same time as, a rational, problem-solving approach, the encouragement of science, and its application to economic growth. Successful industrialization seems to require acceptance and internalization of such attitudes, not only by the entrepreneurs themselves, but also by the labour force. Each must master and accept as legitimate the routines of the factory and the bureaucracy, and the relatively long training which is a prerequisite for most positions, and especially those with high status, in an industrialized society. This is associated with a set of values which judge people by what they can do, rather than by the social status of their parents or their physical features; which expects them to treat every client in the same manner, disregarding the other social roles which that client may play in other situations; which expects

---

[21] Weber, M. *The Protestant Ethic and the Spirit of Capitalism*. (Eng. trans.) Scribners, 1930.

business and other social relationships to remain free from emotional exchanges or any show of strong feelings; which expects the obligations of the relationship to be relatively precise and limited to the matter in hand; and which stresses that the individual's primary responsibility is to himself rather than to the groups with which he happens to be associated at a particular time. This last expectation means that, if a choice has to be made because the organization for which one works is declining, the individual should seek another job rather than remain with the organization in the hope that it can be saved.

This section has commented on some of the necessary conditions for the development of full industrialization. Judgment that a condition is necessary must remain tentative, of course, since one exception would be enough to refute it. What we can offer, then, is a set of conditions which have always accompanied widespread industrialization so far, and whose absence would have prevented it. As in the first section of the chapter, there is little to suggest that any one, or any set, of these necessary conditions forms a sufficient condition. The range of conditions has covered the organization of trade and the firm, the accompanying institutions, and the values of an industrial society. These frequently form a moderately consistent pattern, so different from the patterns of the preliterate or the 'feudal' society that the transition from one to the other represents a veritable social revolution.

Wirth's Theory and the Industrial City

WHILE material on the pre-industrial city was relatively scarce and scattered at the time Wirth was writing, data on industrial cities was already plentiful, though not usually organized to provide tests of his theory. The present chapter therefore begins by selecting some of the data which are most pertinent to Wirth's theory, and attempts to test it. The organization of this part closely parallels the second section of the preceding chapter. In particular, attention will be focused on the structure of the city rather than on the processes within it. These tests will be followed by an examination of the evidence on classificatory schemes for cities; evidence on the social areas within a city; and finally on the functioning of primary groups within the context of the city. These subjects will occupy the four sections of the chapter.

In reviewing some of the available material, it will be necessary to maintain very clearly the distinction between the city defined in terms of size and the urbanized society, defined in terms of its social characteristics. Since Wirth's propositions could be applied to either or both of these, there was no reason to insist on this distinction in the first chapter. The second chapter referred only to pre-industrial cities which are found largely, if not entirely, in societies which are not urbanized. For the present chapter, by contrast, the distinction becomes important, since the results may depend on which comparison is being made. The city in an urbanized society is likely to be very different from the village in a society which is not urbanized; it may be much less clearly distinguishable from a village in the same urbanized society. It is therefore important not to confuse the differences between city and countryside within a society, with the differences between an urbanized society and one which is not urbanized. One would expect Wirth's description of the urban way of life to apply most strongly to the residents and workers in large cities in an urbanized society, and least strongly to rural workers and residents in 'feudal' societies. It is much less readily apparent whether rural workers and residents in an urbanized society will fit Wirth's description better than city residents and workers in a 'feudal' society.

The evidence in this chapter deals mainly with rural-urban contrasts in an urbanized society; where the data relate to contrasts between urbanized and 'feudal' societies, this will be made clear in the text.

I. SOME EVIDENCE RELATING TO WIRTH'S THEORY

The major premise from which Wirth's theory starts is that size, density and heterogeneity are highly significant elements in the definition of a city. It follows that these three variables should be correlated with other distinguishing features of the city, even though they may not furnish a full description; it also follows that unless one is content with a simple dichotomy between city and country, the other significant features should change in a systematic manner as size, density and heterogeneity grow. It is not enough to find differences between the largest cities and the most isolated rural areas; there must also be meaningful relationships between these variables for intermediate sizes of town. The fullest empirical tests of these deductions have been made in the USA, though some additional information is available from the study by Moser and Scott. One of the most elaborate studies was by Duncan and Reiss,[1] using the social data from the 1950 US Census. They took eleven categories of size, ranging from cities with over three million inhabitants to farm families in unincorporated areas.[2] They found that the cities within any size category were, on the average, less densely settled than those in larger size-categories, and more densely settled than those in smaller size-categories. Hofstaetter confirmed this for large metropolitan cities; but in his research the correlation fell below 0.2 for smaller and more isolated cities.[3] It does not necessarily follow that there is a direct relationship, for any given city, between its growth in size and the increase in its average density. Even though 'urbanized (built-up) areas' rather than administrative boundaries were used, the average size and density for each grouping may conceal wide variations. Further, the increase in average size, as one moved from one category to the next, was not closely paralleled by the increase in average density: in some cases, the increase in size was almost twenty times the increase in density; in other cases, only four or five times. Heterogeneity has always been a relatively imprecise term. Duncan and Reiss used three measures

[1] Duncan, O. D. and Reiss, A. J. *Social Characteristics of Urban and Rural Communities, 1950.* Wiley, 1956; Duncan, O. D. 'Community Size and the Rural-Urban Continuum', in Gibbs, J. P. *Urban Research Methods.* Van Nostrand, 1961, pp. 496–514; Dewey, R. 'The Rural-Urban Continuum'. *Amer. J. Sociol.*, vol. 66, 1960.

[2] I.e., towns and villages with no local government powers. Most are in sparsley settled areas.

[3] Hofstaetter, P. R. 'Your City—Revisited: A Factorial Study'. *Amer. Catholic Sociol. Rev.*, 1952.

which might be regarded as signs of heterogeneity: the extent to which the city contained (1) both farm and non-farm employed males, (2) both whites and non-whites, and (3) both native and foreign-born. In each case, there were differences between the size categories, though these were not always very consistent. Occupational heterogeneity, in the sense of farm and non-farm, was marked only in the villages; racial heterogeneity was found throughout the Southern USA, but only slightly more frequently in the cities than in the countryside; while ethnic heterogeneity, in terms of the percentage born outside the US, declined fairly consistently but not very steadily as one examined the lower size-categories. These findings offer limited support for the view that size, density and heterogeneity are related to each other, but not so closely related that one could dispense with one of the three variables in analysing urbanized societies.

We now consider the other side of the claim: that a description in terms of these three key features is a very useful basis for summarizing many different attributes of the city in an urbanized society, and in distinguishing between city and country. Very few studies have considered the correlates of either density or heterogeneity; one is therefore obliged to concentrate on the factors which are reasonably closely associated with size. Two other studies are also relevant here: those of Moser and Scott[4] in England, and of Hadden and Borgatta[5] in the USA. Both studies drew on a wide selection of official statistics for the major cities. Very few of the variables studied by Moser and Scott—which covered selected aspects of population size and growth, population structure, housing, economic character, social class, voting, health and education—showed significant relationships with size. The most prominent correlates of size were the proportion employed in finance, insurance and banking; the proportion voting in the 1955 General Election, which was lowest in the large cities; and to a limited extent the proportions voting in local elections, the size of the poll in absolute numbers, and the rate of TB notifications. The most significant and interesting of these was the concentration of financial operations in the large cities. This has also been found by Hadden and Borgatta, and in the American research on dominance by the metropolitan area over its hinterland, by Bogue,[6] Vance and Sutker,[7] and others. Duncan and Reiss[1] found only a few weak relationships between size and other characteristics of cities in urbanized societies.

[4] Moser, C. A. and Scott, W. *British Towns*. Oliver & Boyd, 1961.
[5] Hadden, J. K. and Borgatta, E. F. *American Cities*. Rand McNally, 1965.
[6] Bogue, D. J. *The Structure of the Metropolitan Community*. U. Mich. P., 1949.
[7] Vance, R. B. and Sutker, S. S. 'Metropolitan Dominance and Integration', in Vance, R. B. and Demerath, N. J. (eds.) *The Urban South*, N. Carolina U.P., 1954.

Fertility was somewhat higher in the rural areas, for both farm and non-farm families; the median number of years' schooling was slightly lower in the rural areas; the proportion of white collar workers was much lower, at least among the farm families; and the proportion of farm families whose members depended heavily on non-farm jobs was much lower in the areas which were farthest from the large cities. Hadden and Borgatta found more variables which were related to size. Some of these were obvious: those phrased in terms of the total number of persons, such as the total size of the city budget, rather than those dealing in rates. Others were less predictable, however; larger cities also had: a higher percentage using public transportation; a higher proportion with incomes over $10,000; a higher median value for owner-occupied homes; a higher proportion of homes declared structurally sound; and a smaller proportion of detached homes. These variables correlated between 0.2 and 0.3 with city size. These authors' reliance on official statistics consistently limited their ability to examine the associations between size and some of the subtler social variables to which Wirth was referring.

While these findings apply only to highly urbanized societies, they do suggest that for such societies the usefulness of size as a criterion is very limited. Although it is clearly associated with statistics in terms of *absolute numbers*, it shows less correlation with those expressed as *rates*. If this conclusion is substantially correct—and there are many types of statistics against which it has not been tested—it suggests strongly that the city differs in scale, but not in content, from the small town within the same society. As Mann found in a simpler comparison,[8] the differences between urban and rural residents have for the most part shrunk to insignificance. The physical differences often remain, but it would be misleading to infer that major social differences lie behind them.[9] This is not to assert that the Londoner and the Shropshire lad have both died out; but that there is so much more exchange between city and countryside that too few 'pure types' remain to show up in official statistics.

Few studies of heterogeneity and its correlates have been made. From one point of view, the large city is more heterogeneous by definition: the greater number of persons means that, even if there are no differences on the average between city and country dwellers, the

---

[8] Mann, P. H. *An Approach to Urban Sociology*. Routledge & Kegan Paul, 1965, pp. 30–65, 69–70. See also Dewey, *op. cit*. Great Britain may well be an extreme case.

[9] A clear example of such a false inference was given in Stewart, C. *The Village Surveyed*. Arnold, 1948. He found a widespread belief that the older homes along the main street housed most of the established residents, while the more recent council houses were full of newcomers. In reality, Stewart's survey found the reverse to be true.

chances that every known group will be found in a city are higher than the chances that every known group will be found in a village. This is supported by Hadden and Borgatta's finding[10] that the economically specialized cities were small, and that economic specialization decreased as size grew. This is not to suggest, of course, that villages are homogeneous. It does imply, however, that size means greater potential in the city for establishing groups. If the village contains only two concert-lovers, while the large town contains two hundred, the town concert-lovers may be much better able to promote their common interests through collaboration. They are more likely to be able to organize formally, to find officers and to organize their own concerts. Their success, of course, will depend on a variety of factors, such as the effectiveness of communication and transportation. The conditions favourable to a formal organization of concert-lovers are also favourable to the formation of other formal organizations, some of which will compete directly for the time and resources of concert-lovers; and the city dweller may experience greater difficulty in discovering who shares his special interests. Duncan[11] tried to express the result of these social forces in quantitative terms, by seeking to establish the minimum size of city or metropolitan area which appeared to be necessary to support particular facilities. He found, for example, that few cities with less than one hundred thousand were able and willing to support a symphony orchestra or a zoo.

This short review has tended to suggest that size may be important mainly in two connections: with the control of wholesaling and especially finance; and with variety of highly specialized amenities. Density and heterogeneity are to some extent associated with size; there is insufficient evidence to show whether they are important in their own right as part of a succinct description of the city. Attention now turns to the twelve propositions which were outlined in the first chapter as the basis of Wirth's theory. Again, the evidence is much less complete than one might wish; but a number of highly suggestive studies have been made.

1. Relatively weak bonds among co-residents; formal social control; physical separation of diverse sub-groups
Wirth supports his point about the weak bonds between co-residents by two arguments. Firstly, specialization in the city has been carried so far, and the residents are so mobile, that co-residents are unlikely to have many common bonds which reinforce that provided by the proximity of their homes. Secondly, he argued that persons with similar characteristics often drifted into the same area, rather than deliber-

[10] *Op. cit.,* p. 70.
[11] Duncan, O. D. 'The Optimum Size of Cities', in Hatt, P. K. and Reiss, A. J. (eds.) *Reader in Urban Sociology.* Free Press, 1951, pp. 768–769.

ately choosing each other, and in other cases they were brought together by a common interest in escaping from particular problems or sections of the city. Wirth's proposition is true by definition if one takes the entire city as the social unit, and interprets it to mean that the ties between two residents of the same city, chosen at random, are less likely to be strong than the bonds between two residents of the same village, chosen equally randomly. On the other hand, if one defines co-residents as those who inhabit a small section of the city, the evidence is much less amenable to simple conclusions. On the one hand, the research by Young and Willmott and by Townsend in Bethnal Green and by Gans in Boston[12] found strong bonds between neighbours; the authors suggested that the sub-cultural tradition, the availability of varied employment in the borough, and the low rate of geographical mobility were important factors. Wirth's own studies of city ghettoes also pointed to the strength of the religious, cultural and familial bonds which united neighbours; at the same time, these bonds did not always prevent many of the children from moving into suburban areas and assimilating when they had the opportunity; nor did they promote non-economic relationships between Jews and Gentiles in the city. Other studies in central areas of large cities, such as Kerr's in Liverpool and Mogey's[13] in Oxford, have placed relatively less emphasis on strong common bonds. Kerr pointed to a relatively clear distinction between members of one's own social group (from whom one should never steal) and outsiders (from whom stealing was regarded as legitimate, though illegal). Mogey stressed both the intimacy of small groups in St Ebbe's and the distance at which the neighbours were kept, even though they were of fairly long standing. Mogey and Young and Willmott also described, in support of Wirth's proposition, the weak bonds which united families in relatively new housing estates; even though in many respects their population was as homogeneous as that of the central working-class areas. It seems reasonable to conclude that co-residence for a number of years is a necessary but not a sufficient condition for the development of common bonds; as Mitchell and Lupton[14] remarked, once the first exploratory contacts have been made, time may be a neutral factor in influencing the strength of particular relationships.

The Berinsfield study by Morris and Mogey[15] examined social

[12] Young, M. and Willmott, P. *Family and Kinship in East London*. Routledge and Kegan Paul, 1957; Townsend, P. *The Family Life of Old People*. Routledge and Kegan Paul, 1957; Gans, H. *The Urban Villagers*. Free Press, 1962.
[13] Kerr, M. *The People of Ship Street*. Routledge and Kegan Paul, 1957; Mogey, J. M. *Family and Neighbourhood*. O.U.P., 1956.
[14] Mitchell, G. D. and Lupton, T. 'The Liverpool Estate', in *Neighbourhood and Community*. Liverpool U.P., 1954.
[15] Morris, R. N. and Mogey, J. M. *The Sociology of Housing*. Routledge and Kegan Paul, 1965.

relationships in an urban fringe area. This, too, was a new housing estate, but forty to fifty per cent of the residents had previously lived together in a slum on the same site. While the Berinsfield families' common bonds as neighbours, tenants and householders were important in the first weeks after the move, their significance as bases for potential joint action was very limited. Parenthood and the need for leisure facilities were important to about forty per cent of the residents; these constituted the most significant shared roles. They did not necessarily lead to interaction, of course; and in some instances their effect was divisive rather than unifying—as in the case where some residents suggested they should cooperate in disciplining other residents' unruly children and adolescents. Research on American suburbs has found 'quasi-primary' or 'pseudo-primary' relations to be common.

Studies of small towns, similarly, have not always presented the picture of stability which emerges from Elmtown and Yankee City.[16] While these studies have emphasized the ability of established residents to rank each other fairly precisely, they have not always made clear that the selection of long-term residents was an elaborate and biasing procedure.[17] At the same time, judgments of people who live 'on the other side of the tracks' have normally been made on the strength of family reputation rather than personal knowledge; and Hollingshead suggested that a youth was likely to be evaluated, in large measure, by the existing reputation of his family as influential, respectable but not influential, or low-class. Further, Lenski's study of a small Connecticut town[18] showed that the long-term residents with extensive knowledge of local reputations may not form the majority; and their evaluations may not be generally known and accepted. Similarly, Stacey[19] found in Banbury that, even among the long-term residents, the 'traditionalists' were quite different from

[16] Hollingshead, A. B. *Elmtown's Youth*. Wiley, 1949; Warner, W. L. and Lunt, P. S. *The Status System of a Modern Community*. Yale U.P., 1942.

[17] Warner has been criticized on a number of counts, of which two are particularly relevant here. Firstly, he mistook interest in making fine social distinctions for knowledge of the town's social structure, and thus tended to rely heavily on the judgment of status-conscious middle-class residents; at the expense of upper- and lower-class residents who were more likely to rely on simple distinctions between 'them' and 'us'. Secondly, he assumed that long residence was a pre-condition for familiarity and accuracy; and hence tended to ignore the judgments of younger residents and more recent arrivals. This necessarily emphasizes the more static elements in the social structure, at the expense of those groups which are least likely to accept traditional evaluations and who have most to gain from social change. See the critique by Kornhauser, R. R. in Bendix, R. and Lipset, S. M. *Class, Status and Power*. Free Press, 1953 ed., pp. 224–255.

[18] Lenski, G. E. 'American Social Classes: Statistical Strata or Social Groups?' *Amer. J. Sociol.*, vol. 58, 1952.

[19] Stacey, M. *Tradition and Change*. O.U.P., 1960.

those who had moved into the town some fifteen years earlier, when the aluminium plant was built. The plant workers and managers, though not necessarily recent arrivals, had more limited knowledge and used quite different criteria in evaluating their neighbours. In villages where 'it takes twenty years before people begin to accept you', the proportion of recent arrivals may be high, and the strength of common bonds correspondingly weak.

Formal social control and the physical separation of diverse subgroups have also received sporadic attention in the literature on urbanism. Clearly, informal control alone becomes increasingly ineffective as the size of the grouping increases, and requires supplementation by more formal methods. These formal methods may not be more effective in influencing the behaviour of those who are members of cohesive primary groups; but they have symbolic and practical significance in so far as they are applied consistently to members of differing primary groups, and to those with few primary group ties. In these situations, and in cases where disputes between members of different primary groups arise, formal control may be the only effective resource for containing the conflict. Such generalizations, however, need to be qualified before they can be very useful: one needs to know how necessary formal social controls are in particular types of situation, and where there are several types of formal control which have diverse effects. In studying voluntary associations, for example, one frequently observes formal behaviour patterns when the association is fairly large, and/or when it contains persons from differing social classes. In this case one tends to find that size and heterogeneity both lead to formalization—not simply as a method of maintaining social control, but equally as a method for achieving goals and adapting to the environment. Again, however, the picture should not be oversimplified: Frankenberg's[20] research in a Welsh village stressed the importance of informal control in small voluntary associations which cut across social class lines. The working class, Welsh speaking majority participated little in the formally-run activities; but their informal discussions behind the leaders' backs, and their group decisions for or against cooperation, had a major impact on the size and success of the voluntary association.

Physical separation of diverse sub-groups is not a sufficient solution to the problem of social control. It may reduce opportunities for conflict, but at the same time it increases ignorance and unfavourable stereotypes about other groups: these make open conflict more likely. Precise comparisons of the extent of residential segregation are difficult to make, from the available evidence. In general, one would expect such segregation to be most marked where there are clearly

---

[20] Frankenberg, R. *Village on the Border*. Cohen & West, 1957.

perceptible racial and cultural differences; and one would expect to find segregation in both the village and the city. Residential segregation in the village, as Mann remarked, is very real;[21] but it may also be less obvious to the outsider because the number of inhabitants is too small for clear 'social areas' to emerge. The importance of religious and class differences will usually depend on their openness to immediate observation; and on the extent to which they are overshadowed by similarities in other respects. The existence of social and ecological processes such as invasion-succession and assimilation tends to reduce the clarity of the dividing lines between social areas, and to introduce overlapping. At any one point in time, the boundaries between social areas are likely to be irregular in shape and erratic; and in terms of social class, for example, only very short streets are likely to be altogether homogeneous. This was the conclusion, for instance, of Collison and Mogey's study of the class composition of the 1951 Census tracts in Oxford.[22] Although the boundaries of these tracts were drawn in a manner which made each of them as homogeneous as could be expected from areas containing several thousand people each, most of them contained representatives from at least three of the Registrar-General's five social classes.

2. *Difficulty of knowing all others personally; impersonality, superficiality, etc.; relationships treated as means to ends*
While it is clearly impossible in a city to know all other residents personally, Lenski's study[18] pointed out that this might not be easy in a town of six thousand. One should distinguish carefully between genuine knowledge and a feeling of familiarity. It may still be possible for established middle class residents of Oxford to believe that until recently everyone in the city knew everyone else; it was still possible after the war for a retired colonel's wife in Banbury to remark to Stacey that there were only three or four families living in that area. Clearly, in so far as these beliefs represented something more than an illusion, they were intended to be prefaced by two qualifications: firstly, 'everyone' referred only to families who were in the same social class as the respondent, and was not designed to include the population in the lower classes; and secondly, 'knowledge of everyone else' did not necessarily refer to personal knowledge, but could include persons linked by mutual friends and acquaintances. The existence of mutual friends served as the guarantee to each that the other was a suitable associate. In addition to those with whom one was linked through common friends, members of higher classes could frequently be identified by name or by organizational affiliation; while those of lower classes could

[21] *Op. cit.*, pp. 72-73.
[22] Collison, P. C. and Mogey, J. M. 'Residence and Social Class in Oxford'. *Amer. J. Sociol.*, vol. 65, 1959.

be placed either by reputation or by their lack of connections with the respondent. Through mechanisms of this type, the range of one's personal acquaintances could be augmented very considerably. Affiliation with voluntary groups or with particular firms increases the range of persons in whose eyes one's status can clearly be established, where personal knowledge is lacking.

The impersonal, transitory, segmental and superficial character of relationships in an urbanized society is again easily exaggerated. Certainly, if one takes two persons at random, the likelihood that they will have an intimate relationship is small. On the other hand, relations in a neighbourhood with little mobility may be complex and at times intimate. Furthermore, the areas where intimacy is commonest may not be the smallest or sparsest. If (as Mann has suggested) friends in a city are more often chosen voluntarily, rather than ascribed by circumstances, friendships in the city may be more intimate than in rural areas of the same size. There is some scope for debate about the nature of roles in the rural areas of an urbanized society, which is relevant to the character of the total relationship between two residents. In a rural area, two persons may meet in a variety of contexts, and play numerous roles toward each other: as members of kin groups, age groups, sex groups, social classes, linguistic groups, churches and voluntary associations. They may also meet as employers or employees and as neighbours. Consequently a person's relationship towards his co-worker is not simply influenced by the roles considered appropriate in the work situation; it is overlaid with the other roles which the two persons play toward each other in differing situations. The pair may cooperate as members of the same sports team, but be rivals as members of different churches; and these latent identities will affect their behaviour together at work. If the two persons are usually members of the same group, they may be united by several ties, and their relationship is likely to be primary. Where they interact mainly as members of opposing groups, their relationship is more likely to be secondary. In either case, it is arguable that the roles played are themselves relatively impersonal, segmental and superficial. They are less often transitory, however; by playing a number of segmental and superficial roles toward another villager, one's total relationship with him is likely to become more personal. From this argument, it would follow that the single roles in an urbanized society are indeed segmental and superficial in either case; but in the village they are less likely to be transitory and impersonal. As a result, one's *total* relationship with another villager is likely to be more primary in nature.

The final point of Wirth's proposition, that relationships in the city are more likely to be treated as means to ends, is perhaps oversimplified. In so far as this part of the proposition relates to motives it falls outside our province here. It is nevertheless relevant in so far as the city-dwellers in an urbanized society expect that others will 'stick to business' in their interaction, and will not regard the maintenance of friendly relationships as an end in itself. Where relations are transitory, they are certainly likely to concentrate specifically on the business to be transacted. The more permanent the relationship, the more each party may gain from making friends with the other and from building up more diffuse obligations. Once loyalty to a person, a shop or a brand becomes a significant element in purchasing, for example, the customer is less likely to seek alternative supplies and suppliers, and more likely to feel guilty if he shops elsewhere or for another brand. Conversely, the customer may expect friendly and preferential treatment, and the absence of 'exploitation', if he is a regular patron. As soon as the relationship is influenced by long- as well as short-term goals, diffuseness tends to enter, because the means which are most effective for achieving short-term profit are unlikely to be the most effective for maintaining and increasing profits over a long period. Furthermore, the nature of the participants' goals may change over time: they may become more interested in maintaining the relationship in a stable form, than in taking measures which might increase its profitability but which also involve risks.[23] City-dwellers may also vary greatly in the extent to which they consciously treat relationships as means to ends: the Machiavellianism studies of Christie and Merton,[24] for example, showed that medical students were consistently more likely than business executives to believe that it was profitable to treat social relationships as means to ends. Finally, one should be careful not to assume too readily that primary relationships are free from exploitative elements: the gains may be less apparent, but one often expects a great deal more in return for one's friendship than for one's temporary patronage. In so far as there is a difference between urban and rural relationships, it may lie more in the opportunities for exercising choice and in avoiding relationships which are not desired, than in the extent to which relationships are treated as means to ends.

3. Division of labour, relationships treated as means to ends; dominance of the large firm; codes of ethics
The part played by city size in inducing people to perceive relationships as means to ends has already been discussed. It appeared that

---

[23] Shopping patterns in Chicago are discussed by Stone, G. P. 'City Shoppers and Urban Identification'. *Amer. J. Sociol.*, vol. 60, 1954. Stone distinguishes the economic, apathetic, personalizing and ethical orientations toward shopping in the city.

[24] Christie, R. and Merton, R. K. 'The Ecology of the Medical Student'. *J. Medic. Educ.*, 1958.

size was of some importance in offering people a wider choice, and in reducing the likelihood that one would play a number of roles toward the same person. The division of labour probably has a similar effect, by increasing specialization. With a given population, this does not reduce the probability that two persons will meet in a given number of role relationships; but it does reduce the probability that these role relationships will be readily compatible and will reinforce each other, leading to a more primary total relationship. The more specialized the role relationship, the less likelihood there is that one party can exactly reciprocate the goods and services which he receives from the other; and the more dependent their relationship is likely to be on either strong personal ties or payment in monetary terms.

The dominance of the large firm, as a condition for stability, is based on the assumption that an increase in the division of labour involves an equal or disproportionately large increase in the problems of coordination. This proposition should perhaps be seen as relating, not merely to firms, but equally to governmental, religious, educational, charitable and other organizations. One can find plenty of examples in urbanized societies of the dominance of large firms. Moser and Scott in Britain,[4] and Bogue[6] and Vance and Sutker[7] in the USA, have noted the heavy concentration of finance, banking and insurance in a limited number of metropolitan cities. Studies of wholesaling by Dickinson[25] and others have pointed to a similar concentration of wholesaling in these cities. Manufacturing, which is less concentrated in a few cities, is nevertheless very often concentrated in a few firms per industry; and their market, like the money market, is international rather than national. In general, one might conclude that wherever the number of articles demanded has been large enough, mass production and the large firms have tended to become dominant in industrial societies.

The position with regard to retailing and services is more complex. Some retailing firms and professionals offering personal services have often been able to resist the trend towards large-scale organization, even where the product may be mass produced or the service highly standardized. Although there are large retailing chains in the food industry, many corner shops survive; and (on the surface, at least) the retailing and servicing of large consumer durables is mainly in the hands of small firms. The survival of small food retailing shops which are members of large chains is partly due to cultural factors. Large shops are geared to bulk purchasing of food, which implies larger, less regular purchases by the housewife, a large vehicle, a large refrigerator, and considerable cash or credit. Where most shoppers lack one or more of these elements, bulk purchasing by housewives is scarcely feasible.

[25] Dickinson, R. E. 'Metropolitan Regions of the United States'. *Geog. Rev.*, 1934.

The combination of giant manufacturing firms and small retailing firms appears paradoxical. This is not the place for a full analysis, but a few suggestions can be made. Large firms, one might suggest, are least likely to develop at a particular stage in the productive process when: the purchaser (whether consumer, next manufacturer, wholesaler or retailer) of the product is relatively ignorant, where the cost of the good or the value placed on the service is relatively high, and where the time between one purchase and the next is relatively long. Goods which satisfy these conditions are frequently termed 'luxury goods'; consumer durables are usually good examples. Under these conditions, choices are apt to be made slowly, salesmen and advisers are necessary to assist or persuade the purchaser, turnover is relatively slow and therefore goods have to be stored yet accessible for considerable periods. Large retailing firms with many small branches have tended to flourish in the sale of 'convenience goods', whose turnover is relatively rapid and selling costs limited; tinned foodstuffs are a good example. Under these conditions, carefully standardized goods and services can be offered more economically by the large firm, which can operate with a lower profit margin on each item.

Codes of ethics for occupational groups are not peculiar to cities in urbanized societies, as we have seen. In such societies, however, they are frequently part of a wider process of professionalization, in which the occupational group also develops an elaborate training scheme, certification of members' competence, and often an employment service. In return, it seeks recognition of its status as a profession, better salaries and working conditions, and tighter control over entrants through higher educational requirements. Codes of ethics, especially with regard to bargaining procedures, strikes, and the demands which may legitimately be made, are also found in trade unions; but unions have not traditionally organized training schemes leading to formal examinations, and their upper age-limits for apprentices have in the past frequently discouraged abler boys from pursuing advanced education prior to being apprenticed. Codes of ethics are also found for social classes, ethnic and religious groups, voluntary associations and businesses. The code of ethics may be important as a basis for integrating a complex group. While specialization necessitates a code of ethics, if it is to be compatible with stability, it also undermines such codes by reducing the number of common interests which the members share and the range of common situations which they face.

4. Division of labour grows with the market; extreme specialization, interdependence, unstable equilibrium
Whenever specialization is profitable, it appears to be axiomatic that an increase in the size of the market will induce a more elaborate divi-

sion of labour. Several examples come to mind readily. One finds that not only do firms specialize in particular products, but whole towns may do so; indeed, most classifications of towns in urbanized societies use the major industry as a principal variable. Urban geographers frequently talk of a hierarchy of functional interdependence, in which the most specialized goods and services are found only in the central city of a metropolitan area; those with a fairly wide market are also found in the smaller towns, while those with a mass market may occur throughout the region. The growing range of occupational skills can quickly be seen from the Registrar-General's *Classification of Occupations*;[26] in 1801 the occupations recognized could be listed on a single sheet of paper; they now require a very substantial book. Similarly, as the membership of a voluntary association grows, its leadership becomes more formal and considerably more elaborate.[27]

Extreme specialization has implied growing interdependence in the economy; its effects, however, are more complex than the phrase 'unstable equilibrium' would suggest. In one respect, this description is very apt: a car production line is a delicately balanced mechanism, whose parts are unusually interdependent; non-cooperation by a small but determined group of men can quickly bring production to a standstill. There is a formidable problem of containing the tensions which build up on a production line, where poor work in one section hampers later workers, and where the speed of the conveyor belt puts considerable pressure on the assemblers to keep working at full speed. In this instance, 'unstable equilibrium' is an evocative phrase. The mutual tolerance of differing groups such as political parties again may represent a relatively unstable equilibrium, which lasts only as long as each group perceives that it has more to gain than to lose from cooperation. While instability is frequently real, one should not mistake the restless and changing nature of the city for basic instability. The city is by nature dynamic in an industrial society, whether it depends on conquest or on technology. Although sections of a particular city may change rapidly, and development plans may quickly become obsolete, the stability of the city *as an enduring form of social organization*, or of urbanized society in general, is hardly in question. Nor should one infer that the large firm in an urbanized society is necessarily less stable than the small firm; the evidence summarized by Caplow[28] showed that in the USA, the bankruptcy rate for large firms was lower than the rate for small firms. The reasons for this finding are complex; and indeed it may apply only if the large firms have a wide range of products. The large firm which mass-produced only

[26] H.M.S.O., 1950, 1960.
[27] Chapin, F. S. and Tsouderos, J. E. 'The Formalization Process in Voluntary Associations'. *Social Forces*, vol. 34, 1955–56.
[28] Caplow, T. *Principles of Organization*. Harcourt-Brace, 1964, p. 34.

one commodity might indeed be highly susceptible to rigidity and subsequently to bankruptcy. When the emphasis in American cars changed from sturdiness to style in the 1920s, the cost of stopping production of Model T's and reorganizing the Ford production line was staggering.[29]

5. *Impossible to assemble all residents together; indirect communication by mass media and special interest groups*

Wirth pointed out that the public for the mass media was anonymous, and did not react in unison; since it did not comprise an organized group with traditions and leaders, the response to mass communication was not comparable with the response to a herald's announcement in a pre-industrial city. Wirth saw the power of the mass media as limited; they could readily bring forth a consensus which was latent; but they had limited potential as a method of developing a consensus.[30] American studies by Katz, Lazarsfeld and others[31] have suggested that small group affiliations have an important bearing on the impact made by the mass media. Their work argued that there were opinion leaders, who were more informed and interested in the subject, and who were relatively influential among their friends and acquaintances. Similar studies of voting behaviour[32] suggested that conformity to the voting of one's peers and family was important; those whose friends and relatives were divided were the most likely to express uncertainty, to change their minds, and to abstain. A number of studies, including that of Morris and Mogey,[33] have found that decisions to buy consumer durables may be made by the neighbourhood group as a whole, rather than by the single family which owns the product. In a similar vein, Young and Willmott[12] observed that the advice offered to mothers by infant welfare clinics was generally accepted only if it met with the grandmother's approval. Finally, the studies of political propaganda[34] have usually concluded that—with the exception of television appearances shortly before General Elections—communication often reaches only those who already agree with its content.

[29] Sward, K. *The Legend of Henry Ford*. Rinehart, 1948, p. 199.
[30] This was a principal theme of his 'Communication and Consensus'. *Amer. Sociol. Rev.*, vol. 13, 1948.
[31] Katz, E. and Lazarsfeld, P. F. *Personal Influence*. Free Press, 1956; Katz, E. 'The Two-Step Flow of Communication'. *Pub. Opin. Quart.*, 1957; Klapper, J. T. *The Effects of Mass Communication*. Free Press, 1960.
[32] Summarized in Lipset, S. M. *et al.* 'The Psychology of Voting', in Lindzey, G. (ed.) *Handbook of Social Psychology*. Addison-Wesley, 1954.
[33] *Op. cit.*, p. 113; Young and Willmott, *op. cit.*; Whyte, W. H. *The Organization Man*. Cape, 1957.
[34] See, for example, Benney, M. *et al. How People Vote*. Routledge and Kegan Paul, 1956, pp. 155–166.

While intimate groups may modify, ridicule or even reverse the impact of communication through the mass media, one should beware of underestimating the possible long-term effects of consistent exposure to a single point of view. Although some of the mass media are relatively impartial in their representation of the viewpoints of special interest groups, others offer a much more biased account. The effects of prolonged exposure through the mass media to a limited range of attitudes have not received serious study; but it would be rash to assume that they are no more pronounced than the short-term effects.

When one examines the influence of special interest groups, several conclusions can be briefly stated. The influence of the special interest group is likely to depend on a number of factors: the monetary and human resources it can summon and devote to influencing a particular bill or section of the public; the saliency of the topic to those affected; the extent to which those affected can be readily organized and will become vocal; and the number of persons affected. The number of persons becomes significant only in so far as they are organized and dedicated; as a result, interest groups which cover large sections of the population, such as consumers, parents or children, have been much less effective than their numbers alone would lead one to expect. One cannot conclude that the freedom to form associations will necessarily ensure that all interests in the society are represented in proportion to their merits, their numbers or even their salience alone. Interests associated with wealthy and cohesive groups are much more likely to be clearly heard than interests of poor and unorganized people. In this connection, formal organization to protect one's interests has been shown by a number of studies to be more compatible with middle class values and behaviour than with working class behaviour and values.[35] Similarly, those interests which are specific and limited may be inherently easier to organize than those which are diffuse; broad 'public interests' may remain largely undefended because no-one is assigned to represent specific aspects of them.

6. Differentiation and specialization increase as density grows.
This proposition was based on the analogy with the survival of organisms on a limited patch of ground or water; Wirth applied it without modification to human behaviour. Lampard[36] has taken essen-

---

[35] Among the most relevant studies are Dotson, F. 'Patterns of Voluntary Association among Urban Working Class Families'. *Amer. Sociol. Rev.*, vol. 16, 1951; Broady, M. 'The Organisation of Coronation Street Parties'. *Sociol Rev.*, N.S., vol. 4, 1956; Slater, C. 'Class Differences in Definition of Role and Membership in Voluntary Associations among Urban Married Women'. *Amer. J. Sociol.*, vol. 66, 1961.

[36] Lampard E. E. 'The History of Cities in the Economically Advanced Areas', *Econ. Develop. & Cultural Change*, vol. 3, 1955, pp. 88–89.

tially the same position, arguing from the historical growth of towns that as the density of settlement grew, there was more opportunity and a greater necessity for specialization, within and between towns. Particularly for the small town within the orbit of a metropolis, he and Bogue[6] both concluded that specialization was necessary for survival. Similarly, Seymour has pointed out that there was a similar specialization in the supply of food to ancient Rome. The immediate environs provided the city's perishable food, while the less perishable food came from distant parts of the empire. It is notable, however, that cities in industrializing nations have much higher densities but considerably less differentiation and specialization than comparable cities in urbanized societies. The pressure of competition for space is often met by specialization; but among humans at least it may also be met by efforts to increase the market for one's product without specializing further. Increased density also means, of course, a larger potential market within a given area. Where mass production is not advantageous, enlarging the market may provide a lasting solution.

Where differentiation and specialization develop in response to increased density or to the efficacy of mass production, they may take several forms. The output of more specialized products or services may avoid direct competition, at least temporarily. Personalized service or production may be tried, with the object of committing the customer to a single brand or trader. The specialization may apply only to the wrapper: one may be able to buy the same product at several different prices under different labels or with different brand names. Finally, competing supermarkets or department stores may specialize by offering different goods as loss leaders, with the object of attracting customers by offering a few striking bargains and then inducing them to buy more than enough other goods to reimburse the shop for its losses on the bargain items. Conventional economic theory has rarely considered the types of competition which occur when the number of producers is small, and the consumers lack the knowledge to choose among them on technical grounds. Riesman[37] pointed out that competition among a small number of large firms was likely to be restricted to small differences: advertisers who wished to appeal to the 'other-directed' would stress that one product was a little better, but not very much better. One consequence of this carefully restricted competition is that no firm's share of the market is likely to be seriously threatened, and potential conflict is elaborately circumscribed.

7. Close physical but weak social contact; readily perceptible symbols
Perhaps the organizations most clearly based on close physical and

---

[37] Riesman, D. *et al. The Lonely Crowd.* Doubleday, 1955.

weak social contact are the motel[38] and the commuters' train, which tend to be found only in the urban areas of urbanized societies. Wirth pointed out, in a later work,[30] that close physical contact among a heterogeneous population requires that relationships remain relatively formal and transitory. Differences will quickly become apparent when formality is dropped; and this will threaten the reluctant mutual tolerance of friction and irritations which characterizes the rush-hour.

The use of readily perceptible symbols in an urbanized society is not necessarily more pronounced than in a 'feudal' or 'folk' society; people may use insignia indicating their statuses even when their identity is not in doubt. The peculiar significance of such symbols in an urbanized society arises when social relationships are transitory and segmental, and hence the readily perceptible symbols themselves become the principal basis for interaction in many situations, and not merely an 'added touch'. A single symbol may largely determine the course of interaction in a brief encounter; and this leaves one with considerable freedom to choose and manipulate symbols of status for one's own ends with a small likelihood of being 'found out'. Young and Willmott[12] contrast strongly the reliance on material symbols in their housing estate with reliance on personal familiarity and connections through friends and relatives in Bethnal Green, where intimate knowledge was much greater and length of residence was much higher.

Urbanized societies also differ from 'feudal' societies in that they require more varied symbols, since more roles are available. The standardized initiation scars or occupational insignia are replaced by many different educational certificates as the symbols of adulthood for the male. Stereotyping still tends to be directed against those who are little known yet threatening; but instead of stereotyping people as residents of the next village, or as possessors of a certain family name, one categorizes them as members of an occupational group or a political party.

The effective manipulation of symbols in an urbanized society is not altogether simple, however. The levelling influence of mass production may drastically shorten the period over which a symbol can retain its connotation as exclusive. This can be seen most clearly with regard to fashions, where the new styles are copied so promptly that only subtle differences in material distinguish the expensive original from the inexpensive development.[39] Symbols of exclusiveness are likely to endure only if their acquisition is a long process: where they

---

[38] Hayner, S. N. 'Hotel Life: Physical Proximity and Social Distance', in Burgess, E. W. and Bogue, D. J. (eds.) *Contributions to Urban Sociology*. Chicago U.P., 1964.
[39] Barber, B. and Lobel, L. S. 'Fashion in Women's Clothes and the American Social System'. *Social Forces*, vol. 31, 1952.

are readily copied, the maintenance of distinctiveness requires a constant and expensive search for the very latest or most unusual articles.

8. Economic competition determines land use; residential desirability is complex; segregation of residential and commercial land uses... ...
Economists and economic historians have often underlined the importance of competition for land use in economic terms; Haig,[40] indeed, developed an elaborate theory in the twenties, relating land use to the balancing of rents and storage costs against transportation costs. While this is a fairly accurate picture of the ways in which firms are *expected* to make decisions, it may not correspond very closely with reality. Studies of relocation from the Midlands, for instance,[41] have sometimes suggested that conservatism and personal factors are important, because decisions involve weighing one factor against another in situations where the monetary consequences are by no means easy to assess. In such instances knowledgeable guesses may be the most rational procedure for choosing; and since much of the firm's equipment is relatively immobile, the costs of relocating may be very significant. Consequently, miscalculations are unlikely to be corrected by a second move unless they prove to be extremely expensive; and relocation in response to a small shift in market conditions will be rare.

Economic competition for land may be nullified when compulsory purchase powers are used, or when zoning decisions restrict the uses to which land may be put. Further, industrial and commercial firms must compete for land with parks, universities, hospitals, schools and municipal buildings. While these may be obliged to buy land at the market price, planning decisions may protect them from unfettered competition with commercial and industrial users; and once the land is bought, they may be under less pressure than a commercial firm to use it economically. Public buildings are under some obligation to use their land intensively, because otherwise as the population grows, their users will have to keep requesting funds for additional land; but in the case of non-commercial parks the idea of intensive land use may be a denial of the purposes for which the park was set up.

Finally, there may be areas, such as the Calthorpe Estates in Birmingham and Beacon Hill in Boston, where private bodies refuse to allow intensive development, even though it might be much more profitable. Clearly such a policy has economic implications, in that it preserves amenities and increases the prices of existing houses. The implications are not altogether simple, because the profitability of the

[40] Haig, R. M. 'Toward an Understanding of the Metropolis', *Quart. J. Econ.*, vol. 40, 1926.
[41] Loasby, B. J. 'The Experience of West Midlands Industrial Dispersal Projects'. Unpub. Paper, 1961; Long, J. R., *Birmingham Post*, 10/12/1959.

area might well be increased by sub-division of the lots. The preservation of the character of an area may thus prove more important for its owners than the maximization of revenues from the exploitation of its resources.

Segregation of residential from industrial uses, and to some extent from commercial uses, has been a primary aim of much planning policy. This is most apparent when the city in an urbanized society is contrasted with pre-industrial cities or the industrializinig cities of Asia.[42] In most Asian cities, for example, the division into residential and industrial sections is scarcely visible; there is frequently a much clearer division between the old and the Western sections of the city, where only the latter has clearly differentiated land use patterns. The segregation of residential areas from workplaces is far from complete, of course; partly because separation has not been an unmixed blessing, and partly because it was not customary in pre-industrial cities or in the early stages of industrialization. In most nineteenth century developments, separation implied the disentangling of industrial and residential property over a considerable area, and the large-scale movement of those sections of the population which were least able to afford to relocate. Finally, definitions of residential desirability in the middle class have been ambiguous in relation to the separation of home and work. Separation of home from work-place may not be compatible with convenience of home to work-place; privacy and natural surroundings may not be compatible with accessibility to city amenities; economical land and houses may imply uneconomical journeys to work.

9. *Competition and mutual exploitation rather than cooperation at work; life at a fast pace; orderly meaningful routines*

The evidence on exploitation is far from clear, and depends on the meaning attached to the word. In a very transitory relationship, there may be exploitation in the sense of hoodwinking the other parties, or at least taking advantage of the misunderstandings which are apt to occur in all relationships. On the other hand, in a primary relationship which neither party is free to leave, exploitation may be more systematic and inescapable; and such relationships are certainly not peculiar to urbanized societies. Again, if exploitation means seeing the relationship as a means to one's own ends, the comments made on the second proposition would apply. If exploitation means holding wages below a reasonable level, or pursuit of one's own interests at the expense of the employer's or other workers', one can find plenty of examples. Exploitation is in one sense more fully legitimized in an

[42] Breese, G. E. *Urbanization in Newly Developing Countries.* Prentice-Hall, 1966, pp. 55–69; Jones, *op. cit.*, pp. 40–45.

urban society, where one is more likely to find relationships which the parties are not expected to prolong unless they are mutually profitable.

The problem can perhaps be clarified conceptually by using concepts from game theory; although this implies expressing the proposition in a form which is not readily testable. One might suggest that the exploitative relationship corresponds to the zero-sum game, in which the total rewards are fixed, and only their distribution among the players is to be determined in the course of the game. In this situation, a gain by one player implies exactly equivalent total losses by the others. Where the revenues of a firm are fixed, bargaining between groups is in this sense exploitative by definition. A non-exploitative relationship would correspond to a non-zero-sum game, in which the total rewards are not fixed in advance; here both the total and the distribution of the rewards are determined in the course of the game. In this situation, judicious cooperation may increase the rewards of some, perhaps of all, players. The non-exploitative element in work relationships should not be overlooked: the introduction of superior capital equipment or an incentive scheme may be profitable for both management and labour.

The impression that life moves at a faster pace in the city is plausible, though in part it may be illusory. The city-dweller may be regularly exposed to a greater variety of stimuli, and new stimuli may strike him at a more rapid rate. The *impression* of speed, however, arises only when he is so unfamiliar with the relationships among these stimuli that his reactions to them become relatively slow.

The more rigid demands of an orderly routine are often the direct result of industrialization and bureaucratic organization. As a result, one may see the contrasts between the social organization of the large firm, on the one hand, and the pre-industrial trader's household, or the small farm, on the other. The routines of the small farm family and its labourers are regulated by the seasons and by the hours of daylight, but rarely by the clock. Similarly, according to Sjoberg, the merchant in the pre-industrial city opens his shop irregularly, and does not observe unchanging hours or respect punctuality. Williams' study of Gosforth[43] illustrates admirably the striking differences between the two routines which co-existed in the village: that of the farmers and farm workers, for whom the demands of the job and nature took precedence over those of the home or of leisure pursuits; and that of the construction workers at the atomic plant, for whom the demands of the job were defined and limited much more precisely, leaving a relatively clear line between work and leisure. Wilbert Moore[44] also

---

[43] Williams, W. M. *The Sociology of an English Village: Gosforth.* Routledge and Kegan Paul, 1956.
[44] Moore, W. E. *Industrialization and Labor.* Cornell U.P., 1951, Part II.

illustrated the dramatic changes of routine which were necessary to make the adjustment from a small farm to a factory.

10. Simple class distinctions break down as the division of labour becomes complex
Since urbanized society contains many groups, which vary in their values and in their ability to confer power, prestige, wealth and authority the resultant distribution of rewards is likely to be very complex. No single criterion of class position suffices; and status inconsistencies are often found, so that a vicar may have relatively high prestige and low earnings, whereas a bookmaker has less prestige but higher earnings. An occupation may have high prestige in the eyes of one group, and much lower prestige in the eyes of another. Some occupational descriptions are so unusual (such as sagger maker's bottom knocker) or so vague (engineer, director) that they defy accurate classification by most people. Status inconsistency may also arise where a person's status in one group is different from his status in another; although the study of voluntary associations has found little evidence that they alter the occupational status patterns markedly. Finally, a person may be able to change his status for limited periods by moving to a different setting: Tom Harrisson[45] argued that one of the satisfactions of holidays by the sea for many people was the opportunity to cast off their usual roles and to adopt temporarily those of a higher social class.

One should again be careful not to assume too readily that the class structure of the small town or village in an urbanized society is simple. Stacey's research at Banbury[19] found two status systems, that of the 'traditionalists' which was based on family history, and that based on occupation and income at the aluminium factory. Williams[43] found eight social classes in the small village of Gosforth, using the methods of Lloyd Warner;[17] it is possible that these represented two hierarchies, one among farmers and farm workers, and one among building workers, rather than a single pyramid. If Sjoberg's account of the class structure of the pre-industrial city is substantially accurate, it stands in striking contrast with the complexity of stratification in an industrial society.

11. Group loyalties conflict; geographical and social mobility; sophistication
Wirth regarded the city as a place of ferment, where the meeting of different cultures led to secularization and disenchantment with absolute faiths; a place where traditional authority was challenged and

---

[45] Harrisson, T. 'Notes on Class Consciousness and Class Unconsciousness'. *Sociol. Rev.*, 1942.

evidence demanded. The conflict of group loyalties was said by Wirth to arise from the city-dweller's ties with a variety of groups which were not closely interconnected; by implication there was much less conflict in a rural area. There is little reliable evidence about conflicting group loyalties in pre-industrial cities; the evidence from urbanized societies is more complex than Wirth supposed. Conflicts of loyalty appear to be characteristic of all stable societies. One finds them, for example, in those societies which have strong patrilineal clans and strict rules of clan exogamy: a person's male relatives are all members of the same clan, but his female relatives and his wife all belong to different clans. In a situation where descent is strongly traced through the male line, the man's strong loyalty to his clan is to some extent counterbalanced by his relationships with his female relatives, who belong to other clans. These marital links between the clans act as restraints on hostility between them. Similar conflicts of loyalty were found in Frankenberg's study of a Welsh village.[20] Here they were important, not only in restraining the villagers' zeal for the interests of one group at the expense of others, but also in dissuading them from taking decisions or making suggestions in voluntary associations which might arouse dissension. Conflicts of loyalty in the city may be less acute, because there is less interaction between the various groups to which a person belongs; at the same time, the greater number of separate roles which one plays provides more opportunities for potential role conflict.

Geographical and social mobility are often believed to be higher in the city, but the evidence is again far from clear. In most urbanized and in many industrializing societies, there is a large net migration from rural to urban areas; this gives the possibly false impression that the urban areas (which contain many recent arrivals) have high mobility rates, whereas the rural areas (which contain few recent arrivals unless they include commuter villages) have low mobility rates. At the same time, it may well be true that people move more readily from one town or city to another than from one rural area to another. Mobility increases turnover among members of an organization;[46] while this may deprive it of stable leadership, it can also bring new members and new sources of ideas and enthusiasm. Mobility may not weaken local ties to the extent that some writers have supposed: Litwak, Whyte and others[47] have argued that, at least in the USA, lower and middle range business executives are strongly encouraged to take active roles in the leadership of local voluntary

---

[46] Rossi, P. H. *Why Families Move*. Free Press, 1955, Part I.

[47] Whyte, *op. cit.*; Litwak, E. 'Voluntary Associations and Neighborhood Cohesion'. *Amer. Sociol. Rev.*, vol. 26, 1961; Pellegrin, R. and Coates, C. H. 'Absentee-Owned Corporations and Public Policy', in Gouldner, A. W. and Gouldner, H. M. (eds.) *Modern Sociology*. Harcourt-Brace, 1964.

associations, even though their stay in one area may be limited to two years. Since mobility and conflicts of loyalty have not been clearly established as distinguishing characteristics of the city, it would be premature to conclude with any confidence that sophistication also is more widespread in the city.

12. Levelling influence of mass production; pecuniary nexus
Within urbanized societies, the levelling influence of mass production is by no means peculiar to the town. The increasing difficulty of maintaining clear class distinctions has already been discussed; similarly urban-rural distinctions have tended to break down. This is especially true for the non-farm rural families, as Williams[43] made clear. The break-down of urban-rural differences has been furthered by bureaucratization of many services which are not readily amenable to mass production, and by the commercialization of household services which were formerly undertaken by the housewife. The home-cooked dinner faces competition from the restaurant, the self-service cafeteria and the vending machine; the solo window-cleaner or chimney-sweep from the large firm with more manpower and more lavish equipment; and the almost irreplaceable maid and butler are slowly being supplemented by the housecleaning and banqueting firms. Commercialization of personal services reduces the dependence of worker and householder on each other; it reduces both the intimacy between them and the embarrassment which can arise when one party is critical or wishes to terminate the relationship. Such developments tend to encourage two elements of the ideology associated with the pecuniary nexus: the beliefs that one can obtain almost any imaginable service or good, provided only that one is willing to pay sufficient for it; and that if one can legally earn a living by providing a particular good or service, one's product needs no further moral justification.

Bureaucratization tends to reinforce these effects, by spreading throughout the society. This will be discussed more fully in a later chapter; here we note simply that middle class professional associations in school and university teaching, for example, have moved appreciably closer to the traditional pattern of working class trade unions since the Second World War. Meanwhile the traditional unions are gradually becoming more bureaucratized; not only in the sense that the official who desires re-election is unlikely to be seriously challenged, but also in the sense that the unions are becoming increasingly dependent on full-time specialists—lawyers, negotiators and accountants, for example—who are valued even though they may never have had experience as craft members of the unions for whom they work.[48]

Clearly mass production does make possible the manufacture of

---

[48] Wilensky, H. L. *Intellectuals in Labor Unions*. Free Press, 1956.

identical objects. Nevertheless, the *impression* of mass production may be more apparent than its reality. The range of materials for making goods is now far wider than ever before, yet the air of sameness may be more evident than in earlier times. The Dagenham housing estates have had for most visitors an air of utter monotony; yet according to Terence Young[49] over ninety variations in house design were used. Mass production of services is equally possible, in so far as they can be sub-divided into constituent elements, they are needed fairly infrequently, and the people requiring them can be fitted easily into the categories which the bureaucracy has established. In such an environment, if the officials can be motivated to perform their tasks conscientiously, bureaucracy is indeed most efficient. In an environment where tasks are not easily sub-divided, where regular contact between official and client is necessary, and where its categories fit poorly, bureaucracy is much less efficient.

These few comments on bureaucracy will be expanded in the fifth chapter, where it will serve as a principal example of the institutional forms of an urbanized society. The present chapter now turns from its short evaluation of Wirth's theory to examine the evidence for classifications of the city and its constituent social areas.

II. CLASSIFICATORY SCHEMES FOR CITIES: THE EVIDENCE

All schemes for classifying the cities in urbanized societies have had to resolve the problem of devising categories which will be clear and unambiguous without becoming over-simplified. Cities change relatively rapidly, and there are few firm boundaries between one type of city and another. Two approaches have been adopted in recent years. In the first, the primary objective is to give a brief description of each city in terms of its most outstanding features. This approach has been developed by urban geographers such as Pownall and Nelson;[50] in seeking the most outstanding features, they have usually concentrated upon economic attributes, with the object of building up a description of the town's various economic functions. These descriptions have had the merit of assuming that a city may have several significant economic functions; and they also reveal the respects in which very few towns are abnormal. At their best, these classifications have shown both the industrial and the occupational groupings which were relatively over- or under-represented, and also the extent to which each was over- or under-represented, by comparison with other towns of similar size.

---

[49] Young, T. *Becontree and Dagenham.* Pilgrim Trust, 1934.
[50] Pownall, L. L. 'Functions of New Zealand Towns'. *Ann. Assn. of Amer. Geog.*, vol. 43, 1953, Nelson, H. J. 'A Service Classification of American Cities'. *Econ. Geog.*, 1955; also Gillen, P. B. *The Distribution of Occupations as a City Yardstick.* Kings Crown, 1951.

This shows a considerable degree of sophistication, when contrasted with such simple labels as 'university town' or 'steel town'. They are heavily dependent, of course, on the assumption that the pattern of economic specialization in a town is one of its most significant features. This assumption is extremely plausible in some instances: towns which are dominated by expanding or contracting industries, by technologically simple or highly automated industries, by international or very localized industries, may differ widely in the kinds of skill and training their residents need, the rewards which they can offer, the availability of money for amenities, the age and sex composition of the labour force, the types of leisure facilities which are attractive, the age and condition of the housing, the size, composition and growth of the population.

The second approach has been to correlate a selection of statistics for every city in a country, seeking a few complex factors which would account for most of the variation found, and into which many variables could meaningfully be condensed. One may wonder whether such factors can be discovered, whether they are susceptible to several interpretations, and whether the range of statistics included is sufficient to reveal all significant factors; these, however, are questions which can be answered empirically. Two major recent studies have adopted this second approach; their results shed light on the usefulness of economic functioning as a basis for describing and classifying towns. Moser and Scott[4] used almost all the readily accessible official statistics on British towns; these covered population size, growth and composition, health, education, voting, social class, housing, and 'economic character'. This last was a crude breakdown by the percentages employed in services, financing and manufacturing respectively, and not a detailed breakdown by industry or occupation. Their objective was not to test a theory, nor to compare a number of theories, but to devise a classification which would group together all towns with similar characteristics. The four most prominent factors which emerged, *from this particular set of statistics*, varied in their identifiability. The most important single factor proved to be the social class composition of the town; this was quickly identifiable from the pattern of correlations. The fourth most important, housing conditions and overcrowding, was also fairly clear. The other two factors, which the authors saw as most pertinent to population changes between 1931 and 1951, and between 1951 and 1957 respectively, were much less clearly defined. They were susceptible to a number of different interpretations. The Moser and Scott study classified British towns into fourteen groupings on the basis of these four factors, leaving two towns—London and Huyton-with-Roby—which were quite distinctive. The number fourteen is somewhat arbitrary: for other purposes the fourteen groupings could be combined or sub-divided. Two conclusions pertinent to economic function emerged. Firstly, quite a few of the groups were dominated by towns with a special type of economic function: the seaside resorts, the railway towns and the administrative towns tended to cluster together. This was remarkable in view of the crude nature of the economic indicators which had to be used. Secondly, however, only two of these groupings were pure: the seaside resorts and the exclusive residential London suburbs. In the other cases, there were always exceptions: towns which are traditionally assumed to be dominated by one industry, but which fell into the category dominated by another. Sheffield, for example, had the characteristics of a railway centre, more than those of a metal manufacturing town. This is not altogether a drawback, of course: it may dispel inaccurate popular notions, and may lead to further research which elucidates why some towns do not fall into the expected categories.

Equally striking, however, was the evidence of differences between the towns of the North and West and those of the South and East. The 'industrial towns' were heavily concentrated in the North and West, the 'suburban-type' towns in the South and East. Perhaps the terms 'industrial' and 'suburban-type' are misleading. Although the Northern cities do not have such extensive suburbs as London, the South-East is obviously not devoid of industry. It might be more appropriate to suggest that the older, and heavier and extractive industries are concentrated in the North and West, while newer and lighter manufacturing is concentrated in the South and East.

Hadden and Borgatta[5] have extended Moser and Scott's work, using data from the 1960 US Census. Their data were even more diverse, covering local authority as well as Census and other official statistics, and absolute numbers as well as rates. Since their range of statistics was even wider, the number of factors needed to account for most of the variance among them was much larger. Indeed, when twelve factors had been drawn out, considerable variance remained to be accounted for. The problem of attaching a clear meaning to the factors again presented a problem, and some were open to several interpretations. Several conclusions emerged which are especially relevant here. Firstly, size proved to be of some importance, as has already been indicated. Secondly, differences in economic function proved to be significant, but not the most significant, explanatory variables. The industrial distribution was an important element in several of the factors; but these were usually factors of only moderate importance, and the industrial distribution was not closely associated with the three most important factors. Thirdly, it was noticeable that three or four aspects of the industrial distribution were each important in their own right; hence the analysis gave rise to a classification by industry or economic function in terms of the proportions engaged in (i) education, (ii) wholesaling, (iii) retail distribution, (iv) manufacturing and (v) manufacturing of durable goods; these comprised separate and somewhat independent variables.[51] Hadden and Borgatta themselves concluded that economic function was of little significance, as a variable which was closely related to many social variables; while this is perhaps an under-estimate of its significance, it certainly did not have an important bearing on the first three or four factors. These main factors were not always open to straightforward interpretations; the interpretations assigned to them by the authors were: I. Socio-Economic Status Level; II. Non-White Population; (IIa. Extent of Air Conditioning); III. Age Composition; IV. Educational Centre (% living in group quarters and dormitories, % employed in education). Minor factors included Residential Mobility, 1950-60; Population Density; % Foreign Born; and Total Population.

These two research studies enable us to examine the usefulness of some of the other classifications which have been proposed. While Weber's distinction between patrician and plebeian cities[52] was intended to distinguish among pre-industrial societies, it is possible to apply it also to urbanized societies. Moser and Scott gave some relevant evidence when they examined the voting data; this was one of the elements in their largest factor, social class composition. This is pertinent to the locus of political power, in so far as the two major parties can be identified with different social classes.

Harris and Ullman's[53] distinctions were not used explicitly in either analysis. Firstly their concept of specialized function has already been discussed. Secondly, the percentage engaged in transportation has been used, but this is an imperfect indication of a town's importance as a transportation centre. Thirdly, their concept of a central place has two connotations. Centrality in a physical sense has received little attention; but centrality in the sense of domination over a hinterland has been more extensively studied. Some of the statistics on the concentration of financial resources are relevant here. Moser and Scott and Hadden and Borgatta certainly found a fairly heavy concentration in the largest cities; as a result, size was significantly correlated with the proportion engaged in banking, insurance and finance. Dickinson,[25] Bogue,[6] and Vance and Sutker[7] similarly found a heavy concentration of wholesaling and financial control in a few US cities. Smailes, Green, Dickinson, Bogue and other urban geo-

---

[51] The word 'somewhat' is used advisedly here. The factors themselves were independent; but here we are referring to the original variables which were most closely correlated with each factor. For these, the intercorrelations ranged from +.23 to −.21.
[52] Weber, M. *The City*. (Eng. trans.) Free Press, 1960, pp. 149-163.
[53] Harris, C. D. and Ullman, E. L. 'The Nature of Cities'. *Ann. Amer. Acad. Polit Soc. Sci.*, No. 242, 1945; Ericksen, E. G. *Urban Behavior*. Macmillan, 1954, pp. 126–134.

graphers[54] have studied towns, and the role of some towns as central places. It is useful to conceive of towns as related in hierarchies; though the relationships, and especially the boundaries of a town's 'sphere of influence', may depend considerably upon the criterion of relationship which is used.[55] Bus routes, newspaper circulation and retail shopping, for instance, may show very different spheres of influence.

The discussion has already covered three of Wirth's variables for classifying cities. The usefulness of the other three—industrial balance, age and prosperity—has not been tested much. Age and prosperity entered indirectly into Moser and Scott's analysis in two ways: growth rates over the periods 1931–51 and 1951–57 *sometimes* served to distinguish the older, declining towns from the recently growing ones; and the quality of the housing often gave clues to the amount and quality of building undertaken during the nineteenth and early twentieth centuries. While these were both imperfect measures, they were key elements in three of the four factors which Moser and Scott found. At the same time, each of the three measures was associated mainly with a different independent factor; there was little to suggest that they were closely related to each other. The relevance of age and prosperity appears to be real, though more accurate measures will be needed before their usefulness can be established unequivocally.

The other classificatory schemes suggested for urban societies were untestable within the limits of these two studies. Hoselitz's terms parasitic and generative,[56] and Karl Marx's slave-owning, feudal, capitalistic and socialist, implied the accessibility of data which were not used by these authors. Similarly, the classifications of pre-industrial cities usually offered few distinctions which were valuable in the present context. We therefore turn our attention next to the usefulness of the classificatory schemes for describing social areas within the cities of an urbanized society.

---

[54] Smailes, A. E. 'The Urban Hierarchy in England and Wales'. *Geog.*, 1944; Green, F. H. W. 'Urban Hinterlands of England and Wales', *Geog. J.*, 1950; Dickinson, R. E. *City and Region*. Routledge and Kegan Paul, 1964; Bogue, op. cit.

[55] The lack of correspondence between spheres of influence is examined at length in Gibbs, J. P. *Urban Research Methods*. Van Nostrand, 1961; and in Hauser, P. M. and Schnore, L. F. (eds.) *The Study of Urbanization*. Wiley, 1965, esp. pp. 405–408.

[56] Hoselitz, B. F. 'Generative and Parasitic Cities'. *Econ. Develop. & Cultural Change*, vol. 3, 1955.

III. CLASSIFICATORY SCHEMES FOR SOCIAL AREAS:
THE EVIDENCE

The pre-war land use theories of Burgess, of Hoyt, and of Harris and Ullman were outlined in Chapter 1, along with more recent work by Shevky, Williams and Bell. Urban geographers, and in smaller measure social ecologists, have made many studies of particular towns, to judge which of these patterns fitted best. Emrys Jones[57] and others have pointed to other geometrical patterns which might be found. It has been recognized, of course, that these models are abstract and oversimplified: Jones points out that where a geometrical pattern is applied without regard to the features of the site, as in San Francisco, the result 'approaches the ridiculous' because 'the grid is draped over the hillsides and results in the most incongruous steep hills'.[58] These geometrical patterns are usually modified, not only by such features as the location of hills and rivers, but also by the distribution of industry, the availability of transport and transportation routes, and by cultural factors. These abstract designs are essentially representations of the patterns which would result if location was determined solely by economic competition for plots of land, which differed only in their location relative to the city centre. It is relatively easy to show that none of these patterns fits all cities well; it is much more difficult to offer a more viable explanation for the divergences which would be applicable to all cities within a society.

The United Nations study of urbanization in 1957[59] found evidence of only three social areas which could be found in all cities: a modern commercial, administrative and upper class residential centre; an old city nucleus with narrow streets and very high densities; and an area inhabited mainly by recent poor immigrants and possibly squatters. Middle class suburbs and industrial sections were frequently but by no means always apparent; indeed in the pre-industrial cities the upper- and middle-class frequently lived near the city centre.

It has generally been recognized, however, that more evidence to support Burgess, Hoyt, Harris and Ullman may be found if the analysis is limited to cities in urbanized societies. The early comparative study by Davie[60] made a detailed analysis of New Haven, Conn., and used material on twenty other US cities to show that Burgess' hypothesis was so altered when one took account of industrial location that it was of little value. Some of his zones were found to be very heterogeneous; and the towns conformed more closely to a star-

[57] Jones, E. *Towns and Cities*. O.U.P., 1966, pp. 69–73.
[58] *Ibid.*, p. 59.
[59] United Nations. *Report on the World Social Situation*. U.N., 1957.
[60] Davie, M. R. 'The Pattern of Urban Growth', in Murdock, G. P. (ed.) *Studies in the Science of Society*. Yale U.P., 1937.

shaped pattern, in which radiating streets, rivers and railways formed the axes along which different types of development tended to congregate. Commercial development tended to be concentrated in the central area, and to emanate along the major streets, with minor centres at important junctions. Industrial development tended to be concentrated near water and rail transportation, which might be found in any section of the city. Low-class housing areas were readily observable in New Haven and other cities, and tended to cluster around the industrial areas and transport routes. More expensive housing might be found at any distance from the city centre, and in any area which had little industrial development. Mann[61] pointed out that in British towns one could often distinguish a transitional zone undergoing redevelopment, a central business district, and middle class suburbs; but these rarely formed complete belts, and were of very variable width. The outer rings in Britain are complicated by the presence of green belts, and housing estates; in many cases there may be towns of significant size where one would expect the outer ring to be. Commuting in Britain is over much shorter distances than in the USA; this again tends to reduce the distinctiveness of the commuters' zone, except perhaps in the case of London. Finally, a post-war study of the Tokyo conurbation[62] showed a clear division into industrial and residential areas around the central area, and between the major axial roads. The industrial and residential areas were not concentrated in differing parts of the conurbation, and did not fit any simple pattern. The poor did not always reside near the industrial sections; although many did, others lived on the edges of the conurbation, far from their work. While many of the areas which Burgess described are therefore recognizable, they do not often appear to fit the pattern which he described.

In addition to these attempts to derive a pattern for the social areas of a city, there have been more modest attempts to define single types of area and to outline their major characteristics. Detailed analyses of the central business district have been made by urban geographers, and hypotheses have been developed to explain in economic terms why certain types of activity are found at the city centre, while others are dispersed. Haig's analysis of central New York City[40] showed that the competition for scarce land had forced many firms to occupy a number of sites: top administration and finance were located at the city centre, while manufacture and storage were usually located in outlying areas of the city proper. One therefore found specialization, not merely between industries, but within a single firm; only those

---

[61] *Op. cit.*, pp. 76–95.
[62] Ginsburg, N. S. 'Urban Geography and non-Western Areas', in Hauser and Schnore, *op. cit.*

activities which were heavily dependent on a city site remained in the centre.

There have also been many studies of the slum areas, using correlation analysis as the starting-point for explanations of their social organization or disorganization. Many studies in the twenties and thirties showed the concentration of most urban social problems in the slums; crime, delinquency, poverty, overcrowding, sickness, broken homes, child neglect were all heaviest in the same areas. Shaw,[63] indeed, showed that delinquency rates decreased fairly steadily, *on the average*, as distance from the centre of Chicago increased. A number of writers inferred that the economic costs of slums were so high that their replacement would be more than justified by the savings in welfare services which would result.[64] The validity of this conclusion rested on three assumptions; firstly, that the geographical areas which were taken as the units for analysis were significant social as well as physical groupings; secondly, that one could legitimately make inferences from the behaviour taking place in an area to the characteristics of its permanent residents individually; and thirdly, that the physical conditions which were most obviously remediable were the causes of the social and health conditions associated with them. The second and third of these assumptions have been subjected to considerable criticism, from further empirical studies and on theoretical grounds. Lander's more recent study of delinquency[65] found wide variations within each zone, and little evidence of a consistent relationship to distance from the city centre. Robinson[66] illustrated the false conclusions which could be drawn by inferring the characteristics of individuals from those of the groups to which they belonged. Dean[67] delivered a scathing attack on the beliefs that poor physical conditions are the main cause of social problems, and that the abolition of the slums as physical entities would eliminate or at least significantly reduce the social problems associated with them.

The social area analysis of Bell,[68] Williams and Shevky has not assumed that adjacent census tracts could be clustered into larger 'social areas', but has taken the single tract as its major unit of analysis.

[63] Shaw, C. R. *Juvenile Delinquency and Urban Areas*. Chicago U.P., 1942.
[64] See the review and critique by Rumney, J. 'The Social Costs of Slums'. *J. Soc. Issues*, vol. 7, 1951.
[65] Lander, B. *Toward an Understanding of Delinquency*. Columbia U.P., 1954.
[66] Robinson, W. S. 'Ecological Correlation and the Behavior of Individuals'. *Amer. Sociol. Rev.*, vol. 15, 1950.
[67] Dean, J. P. 'The Myths of Housing Reform'. *Amer. Sociol. Rev.*, vol. 14, 1949.
[68] See the references in Chap. 1, f. 42 ; also Bell, W. and Boat, M. D. 'Urban Neighborhoods and Informal Social Relations'. *Amer. J. Sociol.*, vol. 62, 1956; Bell, W. and Force, M. T. 'Urban Neighborhood Types and Participation in Formal Associations'. *Amer. Sociol. Rev.*, vol. 21, 1956.

The third assumption has been avoided in their work, since no inferences about policy have been made; the first and second remain, though the objections to the second have less force since the units are smaller and more homogeneous. As Duncan[69] pointed out, however, the census tract boundaries remain constant over long periods, but the social characteristics of each tract may change rapidly; its unity as a social area may become only nominal. Its main strength, by comparison with the use of larger physical groupings, lies in the finer attention to detail which it makes possible. The social areas were classified according to three variables: social rank (occupational and educational levels): urbanization (fertility, proportion of women who are in the labour force, and percentage of dwellings containing only one family), and segregation (percentage of residents who are Negro, or Oriental; plus the percentage who are foreign born whites from countries outside Western Europe). Shevky and Bell[70] have explained the reasoning behind their choice of measures, but it has not been altogether convincing; in particular, the three measures of urbanization do not appear to be closely associated with each other, either in theory or in practice. The other problem which arises in their approach can more readily be resolved as further data becomes available: are the measures sufficiently closely correlated with other social variables so that knowledge of an area's position on these three indices gives a reasonably accurate picture of its other social characteristics?

An alternative approach to map-reading or more detailed breakdowns has been the use of factor analysis to account for land use patterns. Marble,[71] for instance, tested the usefulness of the Burgess and Harris-Ullman theories in explaining land values for all city blocks. The land value did not bear a linear relation to its distance either from the central business district or from the nearest local business centre. The distance from the nearest arterial highway was the only spatial variable which was significantly related to land values. Knos[72] also found that the multiple-nuclei theory fared poorly as an effort to explain land values; the zonal, central-place and sector theories were more useful; a combination of the zonal and sector theories proved the most useful for American cities. Land values in Chicago and in Topeka, Kansas (population 130,000) were related to the sector of the city, the distance from the city centre, and the distance from the city's main traffic artery. One could also identify

[69] Duncan, O. D. Review of *Social Area Analysis. Amer. J. Sociol.*, vol. 61, 1955.
[70] Shevky, E. and Bell, W. *Social Area Analysis*. Stanford U.P., 1955.
[71] Marble D. F. 'Block Land Values', in Garrison, W. L. (ed.) *Studies in Highway Development and Geographical Change*. U. Wash. P., 1959.
[72] Knos, D. S. *Distribution of Land Values in Topeka, Kansas*. U. Kan. P., 1962.

'socio-economic regions' within the city, which had similar land values and similar histories of development. It should be noted that these comparisons relate to American cities, and are couched in terms of land values. Their relative usefulness in predicting other city characteristics, or in British cities, might be quite different. In particular, one would expect the multiple nuclei theory to meet with more success in Britain.

Geographers and ecologists have also related distance from the city centre to population density. In over four hundred cities and conurbations, without exception, there is a fairly regular, though not linear, decline in density as distance from the centre grows.[73] The rate of decline is less in the larger and less compact cities; and it has tended to diminish in Western cities over time, as transportation has been able to cover greater distances in a given time, and as industry and commerce have decentralized. In pre-industrial cities, by contrast, the limited evidence suggests that the rate of decline has remained fairly constant over time. This may be related to the spatial distribution of the social classes. In pre-industrial cities, transportation and communications are poor, and the wealthy tend to live near the centre. As urbanization has proceeded and more efficient transportation and communication have been introduced, many of them have moved to the outskirts of the cities. This process has been documented in a number of South American cities, and has reversed the relationship between wealth and location relative to the centre.[74]

The differences between pre-industrial and industrial cities are very noticeable in this regard. While one can detect social areas in a pre-industrial city, the segregation occurs in terms of occupation, religion, ethnic origin and other attributes of the group. Group members work, live and spend their leisure in the same area, sometimes in the same building. In the industrial city, segregation by social characteristics has by no means disappeared; but it is often less marked than segregation in terms of type of activity. Group members live, work and spend their leisure in different buildings, and quite frequently in different parts of the city.

IV. THE FUNCTIONS OF PRIMARY GROUPS IN THE CITY

The final section of this chapter examines some of the evidence on the place occupied by primary groups in an urbanized society, and especially in its cities. Wirth makes little mention of primary groups,

[73] Berry, B. J. L. 'Research Frontiers in Urban Geography', in Hauser and Schnore, *op. cit.*, p. 419.
[74] Schnore, L. F. 'On the Spatial Structure of Cities in the Two Americas', in Hauser and Schnore, *op. cit.*, pp. 356-368.

as if he regarded them as survivals from a rural past whose status had become unclear. He recognized their existence, in his account of the ghetto and especially in his analysis of the impact of urbanization on the family. He used his propositions to demonstrate the changes which took place in the family; but did not really integrate this analysis into his general theory, by examining the reasons for the survival of primary groups in a setting which appeared to be inimical to them. Yet it is a commonplace of sociological observation that neither the primary nor the secondary group has ever been found in pure form. Wirth therefrom needed to offer an explanation, not merely for the survival, but for the necessity, of primary groups within the city.

One might put forward the argument that primary groups within the city are 'purer' and more specialized than in rural areas or in a 'feudal' society; that their contrast with the secondary relationships of city life is a major reason for their survival and continued importance; and that in their modified form they function to assimilate the newcomer by offering links between the city in general and the groups with which he has identified closely in the past. This argument will indicate some of the respects in which Wirth's theory applies least satisfactorily to the urban family.

The argument that the family, and to a lesser extent other primary groups, have become purer and more specialized rests upon an analysis of the impact of demographic, economic, legal and other changes over the last one hundred years in Western countries. The demographic changes—falling mortality since the early 1800's, and declining fertility over the period 1860–1935—mean that children are now very likely to survive, and a high birth rate implies large families. At the same time, there is now only a small probability that any marriage will be broken by the death of one partner before its offspring become adults. The gradual spread of efficient, safe birth control techniques over the past eighty years, coupled with these demographic changes, has led to a situation in which marriage no longer implies a family, and the man is no longer under such pressure to wait until he can fully support his bride. Rather, the young bride today may expect to work until her husband is in a position to reach the heights to which he aspires. The reduced likelihood of death, and the lower age at marriage, both lengthen the expected span of married life, and make marriage a more enduring commitment; and since the couple has a certain amount of freedom to choose whether they will have a family, the decision to have a family also represents a more voluntary commitment.

Three economic changes have tended to support the demographic changes: the rise in living standards in the West; the growth of employment for women in non-domestic occupations; and the increase in state economic activity and control. These have permitted the couple a clearer choice between higher living standards and a large family; and have decreased dependence on the husband as the sole wage-earner. For some, it has allowed the possibility that the wife may work full-time and pay for help with raising the children, especially when they reach school age. In other cases, it means that the wife may become self-supporting if the marriage is broken up. Legal changes have strengthened this trend, by increasing the chances that the wife can obtain a divorce and alimony, and by removing restrictions on her ownership of property and economic independence. Some of the powerful negative sanctions which formerly played a part in holding marriages together have thus been weakened. Again, these changes have tended to make the commitments to a marriage and a family more voluntary in nature. The welfare protection furnished by the state has supported this trend; at the same time, of course, the growth in the amount of education required for most occupations has left the children dependent on parental support for a longer time than ever. The net effect, again, has been to make the family a longer-term and more voluntary undertaking.

Differentiation and specialization outside the family have had two major effects: to strip the family of some of its functions, and to reduce the number of primary groups which directly compete with it. The rapid growth in the number and range of specialized occupations and leisure activities has drastically limited the family's competence to offer expert training for each child's special abilities. The family's function in training for adult life has gradually been reduced and has become more specialized. Occupational training, and the development of leisure skills, has for the most part been taken over by specialized agencies: the schools, further educational institutes, sports and youth clubs, welfare clinics, doctors and psychiatrists, among others, have each asserted their claims to give much better guidance to the learner in their special fields. The family's function, in the main, has been to provide basic social attitudes and behaviour, to motivate him to strive for the society's major goals, and to assist or limit him in choosing among the means for attaining them. These changes have tended to make the family more purely a primary group, by undermining those activities which involve authority and instrumental leadership, while maintaining and perhaps strengthening those based on affection and expressive leadership. This greater specialization has frequently been imposed upon the family from outside: as other roles have become more transitory and segmental in character, members of urbanized societies have become emotionally more dependent on their nuclear families and have come to *expect* more affection and support from them. To some extent, the family has been able to adapt and meet the changes in its members' expectations; but this adaptation has often proved difficult. Where the husband and wife both reach the dinner table tired and irritable, yet each expects to find relief from the day's tensions in the company of the other, the marital tie may prove unequal to the demands being made upon it.

While this argument has been couched in terms of the nuclear family, it can also be applied to other primary groups in an urbanized society. The voluntary nature and emotional content of primary relationships is perhaps more pronounced than in societies which are not urbanized. Primary groups function to provide warmth, support and understanding, in a society which does not consider these the highest public virtues. Expectations from primary relationships stand in contrast with the secondary relationships which are expected to be predominant in one's public life. One is assumed to be free, in the primary group, to relax and release tensions; one does not expect to have to compete for status and identity; one expects to be secure. While these expectations are not necessarily met, they are highly relevant to the level of frustration which group members feel. This expectation may be extended to include those in physical proximity with one's primary groups: the neighbours, because they are physically close to one's family, are expected not to threaten this desire for a homogeneous environment in which one may relax. Mann and Form[75] have both pointed to this expectation: Form found that even in Greenbelt, Md., a fairly homogeneous suburb of middle and lower middle class government employees, a considerable minority wished that their town was less socially mixed. The second function of primary groups is to contrast with, and complement, the secondary characteristics which are expected to be typical of social relationships in the city.

The third function of primary groups in an urban setting is to assist in the assimilation of recent immigrants from rural areas and from countries which are not yet urbanized. Wirth wrote that the itinerant Jew carried his community with him;[76] similarly, students of traditional Chinese society have pointed to the function of the clan in meeting the needs of the new arrival in a city. Little's study of West African urbanization[77] emphasized that voluntary associations along tribal lines were formed for purposes of mutual protection and played an important part in the process by which newcomers were assimilated. These examples will make it clear that this function of primary groups could be clearly observed in some pre-industrial societies. There is also evidence that it occurs in urbanized societies; Litwak's

---

[75] Mann, *op. cit.*, p. 165; Form, W. H. 'Status Stratification in a Planned Community". *Amer. Sociol. Rev.*, vol. 10, 1945.

[76] Reiss, A. J. (ed.) *Louis Wirth on Cities and Social Life.* Chicago U.P., 1964, p. 88.

[77] Little, K. L. *West African Urbanisation*, C.U.P., 1965, pp. 47–109.

research[78] showed the range of services which the migrant might receive from relatives even in a highly industrialized area, and even though there was no problem of learning the ways of a new culture. The cultural differences between city and country in an urbanized society are generally less marked; consequently the help is more likely to be narrowly confined to the period of 'settling in'. For the recent arrival from a different society, primary groups are especially important in affording protection against the conflicts, the stresses and the strangeness which he meets in learning urban behaviour patterns. Hauser and Schnore[79] outline three ways in which this function is fulfilled: in orienting the newcomer, teaching him how to behave and what to expect, in giving him access to work and leisure organizations; in maintaining the newcomer's ties with his rural or foreign traditions; and in carrying urban patterns back to the villages, reducing the gap between urban and rural society. The transmission of social patterns and ideas is especially likely in those societies where seasonal work in a rural environment is supplemented by seasonal migration to the towns.

While primary groups often function to aid the recent arrival in assimilating to urban ways, it should not be assumed too readily that urban behaviour is quickly learned, and rural traditions readily abandoned or lost. The primary group may act as a bridge between the two behaviour patterns, choosing to emphasize certain aspects of each in a setting which is meaningful but less threatening than secondary group interaction. In a somewhat different context, Morris and Mogey[80] showed how bingo and jumble sales acted as a similar bridge between family and work relationships, stressing the friendliness of the one and the universalistic reward system of the other. Oscar Lewis took a more radical viewpoint, arguing from research in Mexico City that peasant family life and traditional religion might be strengthened when families moved to the city.[81] Breese suggested that this was most likely to occur when the entire family moved together.[82] One might add that it would also be more likely when the family moved to an area peopled largely by other recent rural immigrants.

Primary groups therefore have several functions in the city, and are by no means anomalous survivals. They are important as voluntary groups to which a relatively enduring commitment is made; as havens from the tensions which arise in secondary relationships (al-

---

[78] Litwak, E. 'Geographical Mobility and Extended Family Cohesion'. *Amer. Sociol. Rev.*, vol. 25, 1960.

[79] *Op. cit.*, pp. 226–227.

[80] *Op. cit.*, pp. 70–74.

[81] Hauser and Schnore, *op. cit.*, pp. 494–495. Lewis found little difference between recent arrivals and urban residents of thirty years' standing.

[82] *Op. cit.*, p. 74.

though they create tensions of their own); and in helping the newcomer who is unfamiliar with the city and its patterns of behaviour. While they take a more specialized form than in societies which are not urbanized, they perform vital functions in either instance. In chapter 5 we shall return to examine the functioning of primary groups in bureaucracies and voluntary associations, to illustrate in more detail the part played by primary groups other than the family.

Ecological Processes in the City

THIS chapter examines some of the processes which have been observed within the structure of the city; in some respects these function to maintain the city in its present form, while in others they tend to lead towards social change. The discussion here will be limited to those processes which have conventionally been regarded as ecological; no attempt is made to cover such social processes as socialization, the assimilation of immigrants and culturally diverse groups, social control, social integration, government, communication and secularization. These are of course important, but an adequate treatment would be too extensive for the present volume. (One social process, bureaucratization, will be examined in the next chapter.) Finally, this chapter will be limited to industrial cities, since the data on pre-industrial cities are too scattered to be readily synthesized. Breese[1] has however given a careful analysis of these processes in developing countries, whose cities are making the transition from pre-industrial to industrial, and which currently show some of the characteristics of each.

The ecological processes to be discussed here are concentration (and deconcentration), centralization (and decentralization), invasion (and retreat), succession (and withdrawal). The results of the other classic ecological processes discussed by McKenzie,[2] segregation and specialization, have already been discussed. We are departing somewhat from conventional ecological practice in linking each process with an opposite or complement, and in treating each pair of processes together. The first pair, concentration and deconcentration, are used here to refer to changes in the distribution of the population in space. Such changes may be the result of two processes: unequal volumes of migration between one area and another; and differences in the rates of natural increase through births and deaths, between one area and another. In Western countries, it has generally been found that

[1] Breese, G. E. *Urbanization in Newly Developing Countries*. Prentice-Hall, 1966, pp. 108–116.
[2] McKenzie, R. D. *The Metropolitan Community*. McGraw-Hill, 1933.

concentration in cities generally results from city-ward migration, since city birth-rates have frequently been below rural birth-rates, and city death-rates have tended to be slightly higher.

The second pair, centralization and decentralization, are used here to refer to the increasing or decreasing dominance by the city or its central area over the surrounding metropolitan region, as the number and types of services performed within the region come to be more or less exclusively found at the centre. While concentration therefore refers primarily to population, centralization refers to the location of industry and commerce, and especially to the location of policy-making within industry and commerce. Clearly these two aspects of decentralization will be related in the case where city growth involves a movement of factories, shops and offices away from the city centre, and some delegation of authority from the head office to the suburban branches.

Invasion and retreat arise in a situation where there is segregation of residential from commercial or industrial land, or of differing types of residential areas. Invasion denotes the arrival in an area for the first time of other social groups, or of a new type of land user, who was not found there previously. Perhaps one should limit the term to instances where the new group or user is unwelcome to the established users. Its complement, retreat, is used here to refer to the gradual movement of the established users out of the area. Frequently the invasion represents the arrival of a lower-status group; and the retreat represents the movement of the established users into a higher-status area, perhaps beginning an invasion of the latter. Retreat need not follow invasion automatically; the invasion may be stoutly resisted; the invading family or land user may be persecuted; or properties may be modified to accommodate the new arrivals without displacing the old. In a situation of city growth, where migrants are constantly arriving and seeking the cheapest areas, and where residents value the social homogeneity of their neighbourhood highly, invasions may be a regular feature as people gradually move outwards from the city centre.

While invasion represents a movement of groups through space, Quinn regarded succession as a movement through time within the same space. Succession also represents the final stage of the sequence which began with invasion. With its complement, which we have here called withdrawal, succession marks the domination of the area by the recent arrivals, whose organizations and values have assumed control, and who may be in the majority numerically. One can see this most clearly when the new group has a quite different culture: new types of shop spring up, while some of the old ones disappear; the leadership of local voluntary associations changes hands; the composition of the teaching staff and perhaps the elected education committee is altered and comes to reflect some of the changes which had taken place in the resident population.

1. CONCENTRATION AND DECONCENTRATION

While concentration and deconcentration are movements in opposite directions, they are treated together because the forces causing one often overlap with those causing the other. Writers relying mainly on Western experience have tended to argue that in its 'immature stages' concentration is the stronger tendency in cities; while in the more 'mature' cities deconcentration tends to be the more powerful. The criteria for maturity have not always been made explicit, and the term may indeed be a misnomer. The underlying argument has been stated most clearly by Duncan, Sabagh and van Arsdol.[3] They suggest that when an area of the city is first developed in response to rapid population growth, concentration increases fairly rapidly. As time passes, and land becomes more intensively used, the costs of converting property to provide accommodation for additional users increase; at the same time, the property is aging and losing value. Consequently the cost of increasing population concentration will rise rapidly; and if the city continues to grow, it is likely to be much more economical to develop more open land at a slightly greater distance from the city centre. Only when properties in older areas have declined very considerably does conversion or redevelopment become financially attractive. In this view, a sufficient explanation for the balance between concentration and deconcentration can be found in terms of economic costs and the prices which prospective purchasers are able and willing to pay for a site and its amenities.

Until the thirties, it was generally assumed that concentration was the more prevalent force in Western cities. Not only did population densities decrease fairly steadily as one moved outward from the city centre; but densities were generally assumed to be rising in all areas. This is still the case in many non-Western cities, where densities are rising in all city areas, with the possible exception of publicly controlled housing developments.

To explain concentration,[4] it was argued that employers tended to congregate at sites which were convenient to their raw materials, suppliers and transport lines; and their movement tended to induce the population to cluster around their workplaces. In turn, the influx of population brought new and convenient markets for the employers' products. As long as producers were tied to a limited number of sites,

[3] Duncan, B., Sabagh, G. and Van Arsdol, M. D. 'Patterns of City Growth'. *Amer. J. Sociol.*, vol. 67, 1962.
[4] This account relies heavily on Quinn, J. A. *Human Ecology*. Prentice-Hall, 1950, pp. 95–113.

and unable to move far from them without incurring marked rises in costs, concentration of population and centralization of production were most pronounced. Displacement of agricultural employees and growing opportunities for employment in manufacture and commerce exerted strong pressure in favour of concentration. The central area of the main city thus became the focus of economic activity; as more firms were drawn in, greater specialization became feasible, and profitable; and the centre was thus able to cater for a greater variety of tastes. It became possible, in retailing, to offer more economical goods in the central shops, and to set up more specialized shops which needed a large population to sell their unusual wares. In some respects, population concentration enhanced the natural advantages of particular sites, by providing a basis for the support of more elaborate amenities and a more specialized and highly skilled labour force. The natural harbour became more attractive when it was equipped with highly developed warehousing and handling facilities, a ready supply of skilled employees, and competition among very knowledgeable firms. For long periods, then, population concentration was a feature of most urban growth.

The very attractiveness of amenities and natural advantages, however, led to more intense competition for their use and control. As long as access to these amenities depended on a central site, land prices rose rapidly, and only firms which could exploit the amenities very efficiently were able to survive on central sites. The rise in land values, and hence in property taxes, made it increasingly difficult for families to reside in the central areas, except in luxury flats or cramped and dilapidated older houses. Consequently, population has been moving out of the central areas of Western cities for some thirty years, faster than new residents have been moving in. This movement may have gained impetus in Britain from the war-time evacuations; but it can also be observed more clearly in the USA, where no significant wartime evacuation took place.

Deconcentration has meant both a reduction of the population living in the city centre, and a disproportionate increase in the population living within reach of the centre but outside the city boundary. According to Hawley,[5] this trend was beginning to show up in American figures during the twenties, and by 1940 suburban growth was very rapid. At this date, attention was being given to the possibility of halting the concentration of population in the London area; and the Barlow Commission was preparing its report on future planning for the major British cities. The increasing congestion of the city centre, and its great vulnerability in case of attack, coupled with improvements in methods of communication and transportation between

---

[5] Hawley, A. H. *Human Ecology*. Ronald, 1950, pp. 422 *et seq*.

city and suburbs, have undoubtedly been major factors in the recent deconcentration. Improvements in transport and other forms of communication have lessened the advantages of central sites, at the same time that increasing competition has made them much more expensive. The result has been deconcentration rather than increased concentration, within the city boundaries. No similar deconcentration has yet taken place in most suburban areas; these areas have tended to draw population both from the central city and from the more rural parts of the region.

Deconcentration has tended to follow radial transport routes, as families have moved slowly out from the centre while trying to remain within reasonable travelling time. The areas between these routes have tended to be built up more slowly. In the US, at least, most movement within cities has tended to be short-distance, outward, and to newer, more expensive property. Deconcentration has generally been most marked in the largest cities, and in the more slowly growing cities. In the more rapidly growing cities, there is also rapid suburban growth, of course; but the differences between city and suburban growth are most marked where growth in the city itself is slowest. From observations of this type, some writers have associated maturation with the cessation of growth within the city boundaries; although in view of the arbitrary nature of most city boundaries this is likely to be a poor measure. It may be more appropriate to judge maturity by the cessation of growth within the entire metropolitan area; but this could be interpreted as a sign of imminent decline rather than stable maturity.

While deconcentration has been strongly influenced by employment opportunities, and by the cost and efficiency of transport, ideological factors must not be overlooked. Anti-urbanism has long had a strong following in Western countries, and has stressed the belief that the city is the home (if not the originator) of social problems, while the countryside is the source of hallowed traditions and character-building strength.[6] The physical and economic costs of congestion, the poor housing and the discomfort of transport systems which have long been subject to more traffic than they can bear, have been imbued by many with moral overtones. The very real problems of organizing city life have been emphasized, and rural and suburban problems glossed over. This ideology has provided a ready justification for deconcentration, and perhaps a spur to it; it has also deflected attention away from the problems involved.

In the case of voluntary movement to the suburbs or outlying areas, the decision to move may be based on unrealistic expectations. The

---

[6] Mann, P. H. *An Approach to Urban Sociology*. Routledge and Kegan Paul, 1965, delivers a sustained attack on this ideology.

family may move with the object of obtaining a more reasonably priced home, a larger garden and a more rural setting; but may expect many of the urban amenities and services to be provided. The cost of providing these, in terms of both financial outlay and changes in land use, may be high; the cost of urban amenities, together with the increased expenditure of time and money on transportation, may more than offset the savings from moving to a less central site. The congestion of central streets may come to be matched by traffic delays on narrow winding roads, ill-suited to heavy commuter traffic, or on super-highways whose exit roads have become equally congested with vehicles. The rural surroundings may persist only temporarily, until neighbouring land has been converted to residential or business use. The social homogeneity may gradually diminish, as the property ages and property values in the area begin to decline. As a form of relief from discomfort, a single move outwards may be only a temporary palliative; and if it is accompanied by a desire to have urban amenities, these require either very great expense or the presence of a densely settled area which can support them. Yet the desire to escape from the city, while preserving many of its attractions, persists; and the city often retains its reputation as the source of grime and congestion and as the home of those too poor to move out. This reputation, indeed, is likely to be reinforced by the outward movement of the more prosperous, and the immigration of the poor seeking employment.

Where planning controls are tighter, planned deconcentration has faced related problems. Two of these will be singled out for brief mention here. The first is the problem of coordinating the movement of population with the provision of employment and of amenities. Examples of poor coordination come readily to mind, though it should not be assumed that they represent the best in planning practice. Housing estates are sometimes built quickly, but amenities and public buildings follow only very slowly; employers may be reluctant to move unless there is an established skilled labour force in the new area, while families are reluctant to move until jobs are already available; conflict may arise between the exporting and the receiving local authorities over responsibilities for improving services for incoming firms. It requires imagination to plan the simultaneous movement of jobs and workers, customers and shops, travellers and transport services, etc. Shops in a new area are unlikely to be profitable if they are opened as soon as the first families arrive; on the other hand, they may not be well patronized later if the residents adjust successfully to living without them. Since there is unlikely to be a single ideal solution to this complex problem, there is a great need for more knowledge about the costs and frustrations of different lags in the provision of housing, work and amenities of various types. This is in part a technical problem of discovering how to organize events in the most orderly manner, after the major policy decisions have been made. Policy considerations arise, nevertheless, in determining priorities and in judging how much delay in setting up amenities, employment or housing will be tolerated by the groups most concerned.

The second problem of planned controls, assuming that a satisfactory balance can be maintained between the need to limit undesirable land use and the need to maintain freedom and encourage initiative, is more fundamental. We do not know at all accurately the social and economic costs of different types of solution to the problem of distribution of the population over the available land. A whole series of questions arises; those given below will indicate their scope. What are the losses incurred by restricting the growth of the London metropolitan area? What price must one pay if one builds new towns so far from major metropolitan areas that commuting is effectively discouraged? What is lost by planning that no new town will have a population much in excess of one hundred thousand? How expensive would it be to halt the drift toward the metropolitan areas, and toward the South-East in particular, by improving conditions and prospects in the areas which are not experiencing population growth? Can the advantages of resisting the current trend, towards deconcentration in the cities and concentration in their suburban areas, be made to sound convincing enough to encourage people to change their aspirations and movements?

II. CENTRALIZATION AND DECENTRALIZATION

Centralization and decentralization refer to similar movements of manufacture and commerce, and in particular to the tendency for policy-making to be drawn into or out of the city centre. While the evidence on the location of policy-making is less accessible, the trend towards domination by a few firms in each manufacturing and extractive industry, and the tendency for head offices and financial establishments to be confined within small areas in the large cities, need little documentation here. Increasing division of labour perhaps inevitably increases the premium placed on coordinative activities and necessitates more centralized control. Some of the reasons for this centralization have already been discussed: the traditional natural advantages of a limited number of sites, which have frequently been improved; the profitability of a more elaborate division of labour and of mass production; population concentration, which provides local markets; and the proximity of good communications networks.

Decentralization of industry and trade began at about the same period as deconcentration of population. Some of the reasons have already been suggested: the increasingly high price of central sites, growing difficulty in obtaining sites which met the firm's special requirements, increasingly slow transportation in the central areas, greater difficulty in finding room for expansion, and greater difficulty in attracting an adequate labour force as the shortage of housing near the centre grew more extreme.

There were still very real gains from centralization, as new methods of large-scale production were developed and as research facilities were established. The increasing state intervention in economic affairs also made concentration attractive. As government influence over interest rates, mergers, tax rates, conditions for foreign trade, employment and location policies has become more pronounced, and as government contracts have become a more significant source of company revenue, employers have become more dependent on their ability to influence, or at least to take account of in their planning, the decisions of various central and local government departments. While the heavy postwar dependence of many US firms on defence contracts[7] is exceptional, government action has clearly exerted a great influence on many types of decision which were once regarded as the private preserve of the businessman.

Although the pressures towards centralization are still considerable, as can be seen for example in the newspaper industry, decentralization has been increasingly apparent since the 1930s. Some of the impetus probably came from the depression, which put many of the more marginal firms out of business and strained the resources of most others. Again, improvements in transport and communication, and more recently improvements in information services through the use of computers, have reduced the necessity for the concentration of all a firm's activities on a few sites. The result has been a centralization of information, decision-making activities, and financial control in a few large cities, and a slow decentralization of manufacturing and consumer services. The decentralization has not always spread very far yet: moves from inner London to other parts of the South-East have far outnumbered moves from London to Wales or Northern Scotland, in spite of wartime experience in the relocation of some key industries and services. Since the effects of decentralization have frequently been difficult to anticipate, decisions to decentralize have often been made reluctantly and with only a limited knowledge of the likely consequences.

Decentralization in terms of the spread of firms over a number of sites is easier to document than decentralization of control. Quinn[8] showed the extent of decentralization in space for US manufacturing

---

[7] Mills, C. W. *The Power Elite*. O.U.P., 1956; Cook, F. J. *The Warfare State*. Macmillan, 1962.

[8] *Op. cit.*, pp. 341–343.

establishments up to 1939, and more recently Alma Taeuber[9] has showed the extent to which retailing within US metropolitan areas has been decentralized since 1948. Before this date, sales of shopping goods were much more heavily centralized than sales of convenience goods. More recently, however, shopping goods sales have been decentralized more rapidly than sales of convenience goods. Taeuber's findings thus support the view that convenience goods sales became decentralized first, moving out as the population became less concentrated; shopping goods stores have been moving towards the suburbs more recently, however, as a larger suburban population has been available to support them. Since relocation is extremely expensive even if planning permission can be obtained, shopping goods stores may be reluctant to move out to the suburbs until decentralization has progressed to the point where many customers are no longer prepared to journey to the city centre except on special occasions, and consequently their trade declines. If current trends in American conurbations are a valid indicator, shopping centres containing major department stores are likely to become more common in the more prosperous British suburbs. Some convenience goods stores remain in the centre, either because their products are so specialized that they depend on a conurbation-wide clientele, or because they serve those who work in the central areas.

III. INVASION AND RETREAT, SUCCESSION AND WITHDRAWAL

Although, following Quinn,[10] a clear conceptual distinction has been made between invasion and succession, many writers have treated succession simply as the end-product of the process of invasion, and not as a separate process with different characteristics. Consequently there is insufficient material to justify a separate section on succession. The process of invasion, and the protests which have accompanied retreat, have received considerable attention, perhaps especially the impact of Negro invasion on land values in a formerly all-white area. Hawley has suggested that while the first invasion is clear enough, succession is still an impractical concept for research purposes.[11] No suitable criteria have been proposed for identifying its stages, or for determining when it has taken place. Analyses have tended to degenerate into treating every change as an example of invasion or succession, whether due to the arrival of a different social group, the erection of a new type of building, or the coming of industrial or trading premises

[9] Taeuber, A. F. 'Population Redistribution and Retail Changes in the Central Business District', in Burgess, E. W. and Bogue, D. J. *Contributions to Urban Sociology*. Chicago U.P., 1964.
[10] *Op. cit.*, pp. 306–308.
[11] *Op. cit.*, pp. 322, 402.

for the first time. He argued that it is useful only in large Western cities, where areas may be sufficiently homogeneous for the process to be clearly traced. The analysis of invasion and succession is thus considerably more complex than the determination of changes in land use—which is itself by no means a simple matter.

Where a city is growing, cycles of invasion and succession may be continually taking place, as differing sections of the population and different types of land user compete for space and desire to extend their holdings and numbers. There have been studies by Firey, Hoyt and Shaw and McKay,[12] for example, of succession among ethnic groups in certain sections of Boston and Chicago; and by Firey of the resistance to succession in the traditionally upper-class and historic Beacon Hill section of Boston. An early study by Hoyt[13] examined retreat and withdrawal among upper-class residents of a number of US cities, and their exploitation of new areas. Hoyt argued that once a sector of the city became associated with expensive dwellings, it tended to retain that association for a long period. Initially, expensive dwellings tend to be erected in the sector of the city which is nearest to banks and offices, he argued, yet furthest from industrial and warehouse property. If lower-priced housing surrounds the high-priced area, or values begin to deteriorate, the rich tend to move outward; their retreat is toward higher ground or toward water-fronts which are not used for industry or commerce, or toward open ground where golf courses and similar amenities can be developed. Where possible, their movement tends to be confined within the same sector of the city as their earlier homes. Prior to the 1920s, the area of the rich tended to be situated along boulevards where (according to Hoyt) they could ride in carriages, confident that they would meet only 'the right kind of people'; and whence they could travel quickly into town as the occasion arose. The advent of mass-produced cars wrought the demise of these enclaves, for their new accessibility made them attractive to visitors. Consequently the rich retreated into wooded areas off the major roads, with winding streets and large gardens; or occasionally into luxury flats near the city centre. One is not sure how generally applicable this sketch is. Hoyt's own maps shown conformity to his sector theory only for Minneapolis; but since they do not reveal the location of industry, warehouses, waterfronts, high ground and open country, they are inconclusive.

[12] Firey, W. *Land Use in Central Boston*. Harvard U.P., 1947; Shaw, C. R. and McKay, H. D. 'Social Factors in Juvenile Delinquency', in National Commission on Law Observance and Enforcement. *Report on the Causes of Crime*. U.S. Govt. Printing Off., 1931; Hoyt, H. *One Hundred Years of Land Values in Chicago*. Chicago U.P., 1933.

[13] Hoyt, H. 'The Pattern of Movement of Residential Rental Neighborhoods', in Mayer, H. F. and Kohn, C. F. *Readings in Urban Geography*, Chicago U.P., 1959.

One of the most consistent research interests in this field has been the invasion and succession of Negroes in formerly all-white areas. Quinn, for instance, studied the resistance of differing areas to Negro invasion in Cincinnati.[14] He found that, in general, invasion into a deteriorating area proceeded more rapidly near commercial thoroughfares than farther away from them; older houses with large yards were less often successfully invaded than similar houses with small yards; large blocks of flats were more resistant than detached or semi-detached houses; invasions were more likely to occur in houses which faced each other across the street than in houses situated on adjacent streets which were back-to-back.

Schietinger, Jones and others[15] studied the effects of Negro invasion on house values, to test a number of different theories. One of the paradoxes here is that whites complain bitterly that the arrival of Negroes brings property values down; while Negroes complain bitterly that they frequently have to pay significantly higher prices than whites for strictly comparable housing. In his study of Baltimore, Jones found that Negroes paid up to 75% above the market value for homes, in the immediate postwar period. Schietinger found no support for the idea that the threat of Negro invasion brought prices down automatically; though the threat probably has this effect if most whites try to move out immediately and put their houses on the market simultaneously. His data suggested, rather, that Negroes were only attracted to the area when prices were already falling, and that competition among Negroes for the more desirable homes raised prices temporarily. In the long run, prices were lower only for houses which had declined considerably in value before the Negroes moved in. More recent studies summarized by Laurenti[16] have pointed in the same direction: that race has an effect on property values only in so far as it is translated into greater or reduced supply or demand. Reduced white demand may be more than offset by increased Negro demand; increased supply becomes a factor only if many established residents retreat and try to sell their homes simultaneously. Such a retreat would, of course, reduce the market price of the homes, whatever the reasons for its occurrence.

In its early days, social ecology constituted a distinct theory of social behaviour in relation to the use of space: a theory based on the concept of competition for scarce space. The outcome depended on the

[14] *Op. cit.*, pp. 359–360.
[15] Schietinger, E. F. 'Racial Succession and Changing Property Values in Chicago', in Burgess & Bogue, *op. cit.*; Jones, C. R. 'Invasion and Racial Attitudes: a Study of Housing in a Border City.' *Social Forces*, vol. 27, 1949; Gibbard, H. A. 'The Status Factor in Residential Succession.' *Amer. J. Sociol.*, vol. 46, 1941.
[16] Laurenti, L. M. *Race and Property Values*. U. Calif. P., 1962.

parties' differing needs for space, and on the resources which they could muster in competing for it. This theory has been assailed over the past twenty years on several grounds, most notably because it has exaggerated the importance of economic resources in explaining the results of competition for land use. Economic explanations are inadequate in understanding the persistence of parks, public buildings, historic sites and non-profit organizations in occupying central sites, sometimes due to public subsidization. Economic explanations are also superficial in instances where political decisions have a major effect on the direction of economic growth. The Berlin Wall, and American military operations in Saigon, have had very visible effects on land use in those cities; while these may be mediated through economic competition for land, the basic factors are clearly political. Similarly, religious and educational organizations frequently occupy land which could yield profit to business firms; yet no-one seriously proposes that Westminster Abbey or the London colleges and museums should be moved to the outskirts of the city, and replaced by department stores and office buildings. Consequently only certain organizations are expected to compete in economic terms; and as planning controls are broadened, competition becomes increasingly dependent on legal, political and technical questions. Where segregation is strictly enforced, minority groups are in effect forbidden to compete economically for land which has been designated as outside their ghettoes or reservations. When these factors are added to geographical factors, which vary widely from city to city, explanations in economic terms alone are seriously inadequate. On the other hand, if the original concept of competition is broadened to include simultaneous competition in economic, political, ideological, legal and planning terms, its complexity becomes forbidding and its value as an analytical device becomes more doubtful.

From these observations, the value of the ecological approach might appear insignificant. Although some of its key concepts need modification, this approach has formed the basis for a variety of empirical studies which have given a much fuller picture of the spatial organization of cities. These have pointed to the importance of the spatial component as a limiting, and sometimes a determining, factor in society. They have also stressed very consistently the relevance of *process*, by contrast with the persistent focus on *structure* which has until recently characterized most twentieth century sociological work. These emphases have been valuable correctives to an over-simplified and static view of the city as a series of intangible patterns of behaviour, even though they have fallen short in their attempts to offer a full explanation of city life. The ecological explanation has often paid little attention to variations in the social conditions within which competition takes place, the limits to competition which societies set, or the range of competitive resources available to residents, employers and public bodies. The limits to competition, as we have seen, comprise not only the legal and political restraints which are available; they comprise also those areas of the city whose use is placed outside the field within which competition for land takes place. Finally, they include the limits to competition set up by bureaucratically organized firms themselves: instances where land is acquired for aggrandisement and prestige purposes rather than for profit, or for purposes which have only tenuous connections with the maximization of returns to the investors.[17] The structure and processes of bureaucracies themselves will form the subject of the next chapter.

[17] Firey W. *Man, Mind and Land*. Free Press, 1960, offers a theory which takes account of non-economic factors: see esp. chap. 8.

H

Bureaucracy in Urban-Industrial Societies

THE preceding chapters have discussed in very global terms the applicability of Wirth's theory to societies containing cities. This chapter maintains the emphasis on urban-industrial societies, but moves to a lower level of abstraction, concentrating on bureaucracy as a pattern of organization within a society. We are not concerned here primarily with 'patrimonial bureaucracy', which is found mainly in pre-industrial societies—although the Nazi bureaucracy[1] had many features of the patrimonial type: office-holding was dependent on loyalty to a person; authority rested on the whim of the chief; security and promotion were dependent on favour rather than skill or length of service; and spheres of responsibility were ill-defined, determined mainly by successful encroachment and temporary favouritism.

This chapter, then, is limited to the type of bureaucracy which is usually found in urban-industrial societies. In Weber's classic account,[2] this type of 'pure' bureaucracy had five main features: an elaborate division of labour, with a hierarchy of authority and fixed areas of jurisdiction; full-time qualified officials, whose appointment was expected to depend on impartial tests of their technical qualifications; the specialists were expected to receive and deserve autonomy within their areas of competence; their performance was expected to be independent of their personal feelings, about either the case or the general policy; and finally, they expected and were rewarded with stable careers and promotions, and predictable expansion of their responsibilities. Many writers have used Weber's model as a framework for analysing large organizations. The purpose of this chapter, however, is to use Weber's model mainly for definitional purposes, to limit the scope of the phenomena being studied. The chapter is organized around the twelve propositions extracted from Wirth's theory; it starts with the presumption that if Wirth's theory is valid,

[1] Burin, F. S. 'Bureaucracy and National Socialism', in Merton, R. K. *et al. Reader in Bureaucracy.* Free Press, 1952.
[2] Weber, M. *The Theory of Social and Economic Organization.* (Eng. trans.) O.U.P., 1947, pp. 329-340.

its propositions should hold for organizations such as bureaucracies which are especially associated with urban-industrial society, though they may not be inevitable as an accompaniment to industrialization.

Three points should perhaps be made here for clarification. Firstly, in adopting Weber's definition of a bureaucracy, attention is being specifically limited to the 'permanent officials'. We are not including elected representatives or boards of directors whose role is limited to broad policy guidelines, to fund-raising, to strictly political activity for the organization, and to the function of legitimating the bureaucracy's activities to the rest of society.[3] Secondly, the primary interest of the chapter is on bureaucracies within which employees spend only part of the day. No attempt is made to relate Wirth's theory to the peculiar features of those 'total institutions' such as prisons, boarding schools, isolated military bases and monasteries, within which the person's entire day is spent.[4] Total institutions in urban-industrial societies are frequently bureaucratic, but they tend to have a number of special features: tight scheduling of all activities; a small supervisory staff whose object is to make sure that the rules are followed faithfully; an almost uncrossable status schism between supervisors and 'inmates'; a relatively harsh set of expectations; and an administrative elite which assumes responsibility for all aspects of the inmates' lives. Finally, of course the term 'bureaucracy' is used here to describe a particular type of social organization: it is not intended to carry either flattering or pejorative overtones.

1. Relatively weak bonds among co-workers;[5] formal social control; physical separation of diverse sub-groups
The bonds among co-workers rarely approach the intensity reported by Dennis *et al.* in the coal industry[6] or by Horobin among fishermen.[7] Nevertheless there may be strong informal groups among persons at the same status level in a bureaucracy. Many factors are likely to influence the extent to which informal groups emerge, strengthening and diversifying the bonds among co-workers; while these groups have been widely studied, we know relatively little yet about the factors governing the extent and directions of their influence in general.

[3] Parsons, T. 'General Theory in Sociology', in Merton, R. K., Broom, L. and Cottrell, L. S. (eds.) *Sociology Today*. Basic Books, 1959, pp. 13-15.
[4] Goffman, E. 'The Characteristics of Total Institutions', in Etzioni, A. (ed.) *Complex Organizations*. Holt, 1961, pp. 312-341; Goffman, E. *Asylums*. Doubleday, 1961.
[5] Since we are concerned mainly with the organizations within which people are employed, 'co-workers' is more appropriate than 'co-residents'.
[6] Dennis, N., Henriques, F. and Slaughter, C. *Coal is Our Life*. Eyre & Spottiswoode, 1956.
[7] Horobin, G. W. 'Community and Occupation in the Hull Fishing Industry' *Brit. J. Sociol.*, vol. 8, 1957.

One would expect the development of informal groups, and hence the strength of bonds among co-workers, to be affected by superiors' expectations, by the degree of contact which is necessary on the job, by the size of the grouping of status-equals; by the amount of competition between co-workers; by the length of time which they spend in a common status; and by their opportunities for informal interaction during working hours.

Employers' expectations may range all the way from those bureaucracies which officially discourage informal relationships among employees, or between employee and client, to sales-oriented firms which stress the building of personal and even intimate relations as an essential occupational skill. If the employers' expectations are consistent, they may be strong enough to offset some of the effects of rapid turnover. Whyte's *Organization Man*,[8] for instance, argued strongly that where the firm expected its future executives to be out-going and well-rounded, they tended to conform to this expectation, in their private as well as their public life. It was not clear how far this pattern depended on the initial selection of the more out-going and well-rounded applicants, and how far its continuance depended on its perceived efficacy in bringing promotion to those who conformed most closely. At the same time, the importance of superiors' expectations is limited by their ability to maintain close supervision, a network of 'informers', and an atmosphere of constant insecurity. Where these conditions are maintained, supervisors' expectations are likely to be extremely important. Where the subordinates are able to develop a 'closed network', restricting or totally controlling the flow of information back to the superior, the impact of the superior's expectations is greatly reduced. Competition among co-workers is linked to insecurity; where competition can be maintained, and there is considerable insecurity about its outcome, peer groups tend to be weakened. This is especially likely when the group of subordinates is large, unless there are relatively frequent opportunities for unsupervised interaction.

The practice of grading American students or British A-level candidates on a curve is a very clear example of a policy of divide-and-rule which weakens peer-groups. Given that grades are important to students, this practice of grading according to relative rather than absolute performance places the student in a dilemma wherever informal communication is limited. His narrow self-interest would encourage the student to 'cheat', in the specific sense of collaborating with as many other students as possible to do the minimum work and to hand in answers identical with every other student's. This would both eliminate the anxiety of not knowing one's relative standing, and determine how much work was necessary to achieve a given goal; it

---

Whyte, W. H. *The Organization Man.* Cape, 1957.

would also lighten the work-load very appreciably. There are severe limits to the effectiveness of such collaboration, however. From the point of view of success within the examination system, narrow self-interest implies that the student should pressure other students to keep 'cheating', while he himself surreptitiously does extra work which will ensure him a high relative position in the examination. The teacher's policy of divide-and-rule for marking purposes in this situation depends on the tightness of the student organization, and on his ability to detect cheating among small groups of students. Research on cheating in American colleges has generally found that, while most students disapprove of cheating by others, and are aware that it may deflate their own grades, few are willing to take the initiative in reporting instances of cheating to the authorities.[9]

There are a variety of reasons for reluctance to report cheating: some of these throw light on the workings of bureaucracies. Firstly, the line between honest and dishonest collaboration is by no means easily drawn, and is clear only in relatively extreme cases. Secondly, obligations to friends may conflict starkly with the obligation to report cheating. Thirdly, the student's main goal in an examination is to do his best under difficult conditions; and relatively few are likely to have the time or the desire to watch their peers closely unless they themselves are involved in cheating of some form. Yet the bureaucratic system of justice requires that one's observation be reliable and valid enough to withstand close questioning, and possibly a face-to-face encounter with the accused. Unless the accuser already dislikes the accused, or is deeply shocked by his behaviour, he is unlikely to be willing to face this ordeal. Fourthly, the general expectations of the teaching and student bodies with regard to cheating are important in revealing how seriously proctoring is taken, and in teaching the student how far honesty pays and how far dishonesty is punished. In universities where cheating is widespread, the probability of being punished is less; and for the student who believes that cheating is effective, the chances of attaining a high grade honestly appear to be small. Finally, since the student occupies a subordinate status for a given number of years, he is unlikely to identify closely with his teachers unless he associates with them intimately and unless he perceives identification as a means to attaining their status within a

[9] In some universities, such as the University of Virginia, there is an 'honor system' which not only condemns cheating, but equally condemns failure to report cheating. Exams are not proctored: each student is placed on his honour and is subject to the scrutiny of his peers. While no systematic study of this system has been made, to my knowledge, a few of my students have testified to its effectiveness. This does not necessarily imply, of course, that it would work generally. In addition to the unpublished material of W. Bowers, attitudes to cheating have been studied in Stouffer, S. A. 'An Analysis of Conflicting Social Norms'. *Amer. Sociol. Rev.*, vol. 14, 1949.

reasonable period of time. Although his identification with the student body may remain weak, his lack of identification with those in authority may mean that he has little incentive to report student behaviour, even though he disapproves of it.

This brief analysis of informal groups in a university is intended merely as an introduction to a complex subject. It has been concerned to examine only a few of the factors which commonsense suggests as important in influencing their development. While it is very plausible that each of these may have an effect, it is much less clear what will happen when several of them work simultaneously in different directions: under what circumstances do the expectations of superiors have more influence than the length of time the subordinates have worked together, than the opportunities for informal interaction, or than the size of the subordinate group? Questions of this type require much fuller research and much better theory than the simple, vague statement that superiors' expectations are sometimes important.

From a different point of view, one could begin by defining the problem in more specifically sociological terms. The development of informal groups and common bonds among co-workers will depend on the functions which they can perform. Their structure and functions are related: the number of subordinates and their opportunities for interaction will influence the type of structure which can develop; and this in turn will affect the extent to which the informal group is able to perform certain functions, and the extent to which the subordinates expect their peers to follow certain behaviour patterns. The problem then becomes: under what conditions can an informal group perform certain functions for the bureaucracy and for its own members?

One function which the informal group may perform is leadership. Small-group research has isolated several types of leadership, which include task leadership, socio-emotional leadership, ideas leadership, and organizational leadership.[10] To these we may add leadership in recruiting new members and prestige-giving leadership. In most bureaucracies, task and organizational leadership are explicitly vested in the superior, who has formal responsibility for distributing the work, setting an example and stressing a high level of 'output'. Recruitment is frequently designated to a specialized personnel department; prestige-giving leadership is usually the province of the board of directors or the elected representatives who determine broad policy. Leadership in suggesting new ideas may be formally the province of a research and development or organization and methods section. Where

[10] Bales, R. F. 'Task Roles and Social Roles in Problem-Solving Groups', in Maccoby, E. E. *et al. Readings in Social Psychology*, 3rd ed. Holt, 1958; Etzioni, A. 'Dual Leadership in Complex Organizations'. *Amer. Sociol. Rev.*, vol. 30, 1965.

one of these functions is not effectively performed for a set of subordinates through official channels, informal groups are likely to develop to perform it. This is most likely to happen in the case of socio-emotional leadership, for which most bureaucracies until recently have made no formal provision. Over the last twenty years, efforts have been made, especially in the US, to train executives and foremen in 'human relations'; but most of the ensuing research has been interested in the effects on production rather than on informal groups.

Where no formal provision is made for socio-emotional leadership, this need may be met in a number of ways. One common solution is that this responsibility is informally attached to the person who is second in command. Consequently one frequently finds that the chief in the agency or department is concerned mainly with his subordinates' performance in formal roles, exercises his authority to maintain 'output', and is generally treated with respect or fear rather than affection. Meanwhile his deputy, who is not immediately and personally held accountable for 'output' establishes much more informal relations with his subordinates, and assumes the role of a popular, socio-emotional leader.[11] This solution depends to some extent on the personalities of the chief and his deputy. One sometimes finds bureaucracies where the reverse holds: the chief is the socio-emotional leader and the deputy is the task leader, though this is less common. Another solution is to apply the analogy with the family in the work situation: the chief and his deputy are *both* task leaders, because these are assumed to be masculine roles, while the lady subordinates become the socio-emotional leaders. Again, the success of this depends upon the availability of ladies with the appropriate personalities.

One important function of the peer-group, then, is to 'fill in gaps': to assign someone informally to perform those types of leadership which the bureaucracy does not make available. In the same manner, the group which requires a scapegoat can usually find one informally.[12] Informal groups also function to express the interests of significant status-groupings which would otherwise go unrepresented, and to deal with strains which arise in the work situation. Several examples will make this function clear. Becker[13] noted the strong expectation among Chicago school-teachers that the school principal

---

[11] If he subsequently succeeds the current chief, his behaviour is apt to change in accordance with the new expectations of his superiors, and his subordinates will then complain and ask, 'What's come over Old So-and-so?'

[12] In my secondary school, for instance, every peer group had a 'fat boy' who was neither athletic nor scholastic, and was the easy-going butt of jokes. If no group member was overweight, the boy with the roundest face was 'volunteered' to play this role.

[13] Becker, H. S. 'The Teacher in the Authority System of the Public School', in Etzioni, *op. cit.*, pp. 246–247.

would back them up publicly in disagreements with parents, whatever the merits of the case; and that if the teacher needed to be censured, this should be done quietly 'behind the scenes'. He found that discontent was particularly strong where it was believed that the pupil—directly or through parents—could successfully appeal to the principal to override a teacher's decision. Blau[14] found that government inspectors developed their own informal rules for dealing with attempted bribery by the firms under inspection. Officially the inspectors were required to report any attempted bribe to their superiors, with a view to prosecution. In practice, the inspectors were dependent on the co-operation of firms for the performance of their work. Even where the firm was legally obliged to provide the information, this could be done in a very time-consuming, exasperating manner which would greatly lengthen the inspector's job and make him appear incompetent to his supervisors if he tried to be thorough. Faced with this dependence on the firm's good-will, which came into conflict with their rules as soon as bribery was attempted, the inspectors informally agreed to report actual but not attempted bribes. In this way, the threat of reporting an attempted bribe could be used as a method of maintaining compliance from a reluctant firm.

As a third example, Roy[15] described a factory which had strict rules about the use of tools at work. Since no employee was assumed to need a full set of tools all the time, a number of tools of each type were kept at an issuing centre. The employee was expected to go to the centre to obtain the tools for a particular job, fill in appropriate requisition slips, and return all tools at the end of each job, if not at the end of each day. These regulations were ostensibly designed to promote efficiency; but Roy argued that they had precisely the opposite effect, since the supply and demand for particular tools were difficult to match and frustrating delays resulted. Informal rules were therefore worked out between employees, supervisors and issuers: according to these, and in defiance of the official rules, certain employees were permitted to hide tools for their own exclusive use; and issuing slips were then written to reconcile what had happened with what was officially supposed to happen. Roy argued that the informal rules generally increased efficiency, whereas management efforts to impose tight control over issuance of tools consistently led to friction and evasion, and lowered output.

Three points should be stressed in concluding this preliminary analysis of the bonds among co-workers. Firstly, there is no simple general relationship between the development of peer-groups and 'output'. Some peer-groups deliberately restrict output and 'sabotage'

---

[14] Blau, P. M. 'The Dynamics of Bureaucracy', in Etzioni, *op. cit.*, p. 352.
[15] Roy, D. 'Efficiency and The Fix'. *Amer. J. Sociol.*, vol. 60, 1954.

official plans; others invent more efficient means of achieving the organization's official goals than their superiors had provided. Secondly, peer-groups are likely to form at all levels of the organization; although attention has been concentrated on peer-groups of subordinates, this reflects the ease of studying them on the lower levels of the bureaucratic hierarchy, rather than their exceptional importance at these levels. Finally, the question arises of whether the development of informal groups is inevitable, since we argued that socio-emotional leadership is always necessary. Research suggests that they may sometimes be powerful even among prisoners in solitary confinement, whose opportunities for informal contact are extremely limited.[16] Where their development is inadequate to satisfy those needs of the organization and its members which are not met officially, several possible results may be observed. Sometimes the subordinates occupy their status for such a short period that its inadequacies are met outside the organization, or sublimated by the expectation that one will soon graduate to a higher status. Where this is scarcely credible, the subordinates may respond by exhibiting little loyalty to the organization, and by leaving at the first opportunity. If this is not possible, a third alternative is unorganized 'irresponsibility' with regard to the organization's official goals. Examples of unorganized activity of this type can be seen in some (though by no means all) unofficial strikes, in the deliberate damage of production-line machinery, in student riots or vandalism, in embezzlement, tax evasion, and in avoidable idleness by workers or managers. A fourth alternative, which has been little explored, is the expression of work tensions and frustrations in the home, the pub or the club, among members of other organizations rather than among one's co-workers.

We shall return to these two major functions of informal groups, in filling in gaps and dealing with strains, in subsequent sections of the chapter. Informal groups are often more flexible than formal groups; and when this is the case, they may be a more efficient method of dealing with the unexpected, or with longer-term relationships between a bureaucracy and a particular client or policy-maker.

The second part of Wirth's first proposition relates to the significance of formal methods of social control in the regulation of the bureaucrat's behaviour. One thinks of the bureaucrat as a man governed by rules and precedents, rather than as a man governed by the caprice of his superior. The bureaucracy sets up a chain of command, and aims to divide the work rationally into separate areas of jurisdiction. It devises rules to deal with 'cases'. In design, it is not unlike the complex machine which is set up to process standard items from constant raw materials at the greatest possible speed. It is likely

[16] Cantine, H. and Rainer, D. (eds.) *Prison Etiquette*. Retort, 1950.

to be extremely efficient when these conditions are met: when the clients expect to be treated as cases, to be 'processed' in the quickest and most impersonal manner; where lasting relationships or negotiations between bureaucrat and client are unnecessary; and where the cases fit readily and unambiguously into the stereotyped categories used for decision-making. Under these conditions, one can expect objective administration. The formal rules aim to protect the objectivity of the administrator, and the impartiality with which he is expected to treat his clients. They aim to defend the subordinate against arbitrary action by his superiors, and against unpredictable changes of policy. They aim to ensure that the bureaucrat is judged by his loyalty to the rules, and expect him to be equally loyal to any policy which is determined by his superiors. Lipset, among others, has shown that the expectation of political neutrality tends to be met only when the major political parties do not try to introduce rapid and radical social change.[17]

The existence of the rules also protects the bureaucrat against the client. The client normally deals with the bureaucrats at the lowest status-level, who are allowed very little freedom to waive the rules. As a result, the middle and higher level bureaucrats are insulated from petty pressures from the man in the street. The man in the street typically meets only the 'polite but faceless' bureaucrats. The latter are able, quite conscientiously, to sympathize with the client, even to take his side in complaining about injustice; at the same time, they are usually powerless to help him, and can honestly justify their inactivity by saying in effect, 'These rules are greater than either of us'. Finally, formal control operates in the rules for promotions and other rewards. In the more fully bureaucratized organizations, salary scales and conditions for promotion are not only explicit but are matters of public record. The scope for discrepancies in rewards among persons at a given level, or between one bureaucracy and another related one, are therefore limited.

The importance of formal rules can be exaggerated, however. The rules may be inconsistent with each other, even contradictory; or they may apparently contradict the stated objective of the organization.[18] As the number of rules increases, confusion among them becomes increasingly likely. There may be standardized procedures for action in cases where rules clash, and orders of priority among the rules themselves; but these are unlikely to cover all contingencies. One of the pleasures of the subordinate may, indeed, be to get into situations where such conflicts or inconsistencies arise, and the authority of the bureaucracy can be challenged. Some organizations are very incom-

[17] Lipset, S. M. 'Bureaucracy and Social Reform', in Etzioni, *op. cit.*
[18] Barnard, C. I. *The Functions of the Executive.* Harvard U.P., 1938, pp. 163-171.

pletely bureaucratized in terms of rules. Wilensky[19] argued that, until the last fifteen years, American trade unions have been more like patrimonial than 'pure' bureaucracies. Similarly, Gouldner[20] has noted that the rules in business leave large areas uncovered. They may say little about the matters which are of most concern to the workers: opportunities for promotion and conditions of dismissal, for instance. Merton[21] has pointed to the 'trained incapacity' of the bureaucrat who comes to accept the rules as more important than the objectives which they were designed to attain; and who resists changes in them wholeheartedly.

Gouldner[20] has proposed the generalization that obligations within the rules tend to be more precisely defined as one moves down the hierarchy, and that those in the lowest statuses have the least freedom of action. Conversely, he maintains, relaxation of the same rules tends to be more liberal as one moves up the hierarchy. Those in the higher positions are, of course, more likely to be making and interpreting the rules; the rules by which they are bound are correspondingly broader and give more scope for initiative and interpretation. While this is a plausible hypothesis, it is likely to require some modification. Firstly, it is tacitly assumed that there is a direct connection between position in the organization and spontaneous loyalty to its goals. Not only is the person in a higher status expected to be more 'responsible', in the sense of being better able to discern the organization's goals and to make decisions which implement them. He is also assumed to need much less detailed supervision by rules and by superiors to restrain him from putting his own interests before those of the groups within which he works. Secondly, and in return for these privileges, he is apt to be judged more severely if he deviates from expected standards for his status. One expects higher standards of morality from those who hold high office; one is more apt to be shocked, and the newspapers especially interested, if the person in high office does not behave more socially acceptably than the person in a lower position. Thirdly, and in apparent contradiction to the second point, the deviance of a person with high status is a much more serious reflection on the organization than similar deviance by a person with a lower status, who is expected to be less loyal and responsible. Consequently the organization may try much harder to conceal the mistake it made in prompting the disloyal or deviant person if he holds high office. The punishment

[19] Wilensky, H. L. *Intellectuals in Labor Unions.* Free Press, 1956, pp. 243–258.
[20] Gouldner, A. W. 'Discussion of Industrial Sociology'. *Amer. Sociol. Rev.,* vol. 13, 1948, pp. 396–400.
[21] Merton, R. K. 'Bureaucratic Structure and Personality'. *Social Forces,* vol. 17, 1940; reprinted in Merton *et al., op. cit.*; Burke, K. *Permanence and Change.* New Republic, 1935, pp. 50 *et seq.*

for deviance may be no less severe; but it is much less likely to be public knowledge.

The example of the Chicago school-teachers[13] is especially relevant here. The teacher is expected to observe 'higher' standards than the caretaker, and to feel more sense of shame when these standards are not observed. On the other hand, since the teacher is assumed to be more valuable than the caretaker, the organization can dismiss the unsatisfactory caretaker without damaging its own prestige significantly. It can not dismiss the unsatisfactory teacher without casting doubts on its own judgment. This is not simply a reflection of the difference in political power between the teachers' unions and those for caretakers; it is related also to the difference in social valuation which is placed on the teacher and the caretaker, both by the general public and by the school system.

One final comment may be made on Gouldner's hypothesis. The weakness of formal methods of social control over the behaviour of the higher-status bureaucrat should not lead one to believe that informal controls are equally weak. The professor and the civil servant may have almost complete security of tenure formally, and be subject to dismissal only for gross incompetence or when convicted of serious immorality. There may nevertheless be a wide variety of sanctions which can be informally imposed on the high-status deviant. He may be given the least rewarding assignments; he may be deprived of the rewards which are informally associated with seniority; he may find his proposals regularly ignored or defeated, and his requests for favours denied or seriously delayed. He may be put under pressure to resign, and may be made uncomfortable informally in countless ways. These discomforts may extend beyond his present job to his professional reputation in general, thus influencing his chances of obtaining promotion by moving elsewhere. Similar informal control may be exercised by subordinates. The unpopular teacher, or the teacher who has the reputation of being a poor disciplinarian, is apt to be much more severely tested by pupils than the teacher who is reputed to be popular or to have few disciplinary problems. By such informal control, a reputation initially based on such trivial characteristics as appearance or manner may attain a significance for behaviour which is out of all proportion to its original significance.

The third aspect of Wirth's first proposition, the physical separation of diverse sub-groups, is difficult to document from research. In the broad sense, it is frequently obvious. Differing departments are housed in different parts of a building, in different buildings, or even in different areas, depending on the size of the organization and the specialized activities of its departments. Separation restrains the friction which tends to result from constant contact between competing groups; at the same time, it tends to reduce intercommunication to the point where each group develops stereotyped conceptions of the other, and coordination and reconciliation of their interests becomes more difficult.

Physical separation of employees at different status levels within *the same department* also has important consequences for communication, which will be mentioned more later. Goffman[4] has noted that in total institutions there is usually a clear status schism between the inmates and their supervisors which is difficult to bridge, and almost impossible to cross. Each side has a narrow, stereotyped conception of the other; and each has informal groups and informal norms of which the other side knows little and understands even less.

The extent to which physical distance and physical barriers effectively limit communication within a bureaucracy has been little explored. There are a few sociometric studies of factory and office workers which suggest the importance of physical proximity in informal ties, where there are no major status differences; but in other settings the effects of proximity may be severely modified by expectations and the range of common interests which co-workers share.

2. Difficulty of knowing all others personally; impersonality, superficiality, etc.; relationships treated as means to ends

Again, we take it as axiomatic that as the organization grows in size, the difficulty of knowing all others personally is increased. The comments here are concerned simply with attaching a few qualifications which relate to bureaucracies. Firstly, the bureaucrat may be discouraged from establishing personal relationships with his clients, his colleagues, his superiors or his subordinates.[22] As a result, the inherent organizational difficulty of maintaining intimacy may be rationalized as a desirable goal; and the bureaucrat may be encouraged to treat his clients on a 'strictly business' basis. The consequences of this are clear: on the one hand, it assists in ensuring—though it does not guarantee—that all clients who fall into the same category will be treated equally; in this sense, impersonality tends to be associated with objectivity and impartiality. On the other hand, it ignores those other criteria which are officially perceived as irrelevant, even though they might be judged by others to be legitimate grounds for making exceptions to the rules. In the allocation of council houses, for example, 'relevant considerations' are usually defined to include the physical and medical needs of the nuclear family; social and psychological needs, or the interests of groups larger than the nuclear family, are rarely taken into account. A strict interpretation of the rules is liable to break up extended family units, where these persist;

[22] Argyris, C. *Organization of a Bank*. Yale Labor and Management Center, 1954.

and to penalize those families who improve their household amenities and maintain their homes most carefully.

The difficulty of knowing all others personally applies much more obviously to the routine clerk dealing with thousands of applicants than to the higher official, interacting regularly with several superiors, perhaps half a dozen equals, and say twenty subordinates. It also applies with limited force to the central government official who is responsible for supervising the records and safety precautions of a limited number of firms, or for overseeing the implementation of certain laws and administrative regulations by a limited number of local authorities. As Barnard[23] has repeatedly asserted, administration is always dependent on the cooperation of subordinates; and the official who is able to maintain personal relationships with his subordinates may be in a much better position to judge the extent to which, and the way in which, his instructions will be carried out. Thus although the chief may not know all his subordinates personally, and may be careful to leave his immediate subordinates discretion in the execution of his instructions, the bureaucracy may be linked by informal ties which permit efficient communication between its various levels.

The second and third parts of this proposition will be treated together here. Except in cases of obvious exploitation, 'treating relationships as means to ends' is an imprecise phrase. One possible interpretation, in the present context, is to identify it with impersonality, as a situation in which the rules are treated as more important than the person. The importance of formal rules as a mechanism of social control has already been treated at some length; the emphasis here will be more on the factors inducing or resisting the use of impersonal criteria as a basis for judging behaviour. Two points are worth making as a prelude to this discussion. Firstly, although the expectation of impartiality among officials is fairly basic to bureaucratic organization, it may be much more effective in inducing loyalty to the organization and its rules than in creating efficient administration. Kingsley,[24] for example, has argued that officials are rarely impartial on policy proposals, except with regard to trifling matters; their loyal administration of policy depends heavily on the strength of their commitment to its objectives, and only in small measure on their political impartiality. Secondly, while impersonality is much more easily maintained when contact is limited, it can be maintained in spite of contact over a sustained period if the contact is surrounded by an elaborate set of rituals which stress social distance and respect at the expense of intimacy.

Impersonality in the judgment of the bureaucrat's behaviour will

[23] *Op. cit.*, pp. 163–164.
[24] Kingsley, J. D. *Representative Bureaucracy.* Antioch, 1944, pp. 264–278.

be examined here from several points of view: from the point of view of the employer; from that of the colleague; and from that of the client. 'Pure' bureaucracy, in Weber's definition, hires full-time qualified officials, and has no room for the amateur. Appointments and promotions are expected to be made on the basis of impersonal tests of competence in a special area which has been defined as relevant. The organization in this way aims to protect itself against charges of nepotism and 'sponsorship', which are assumed to be both morally undesirable and inefficient in the long run. The question of moral desirability is not at issue here; the subject of efficiency will be raised briefly later.

Two significant questions arise in connection with this account of expected behaviour. Firstly, there may be no relevant technical qualifications for judging the new appointee, or those which are available may discriminate too finely or too weakly to be useful to selectors. This situation has often been met, over time, by a process of 'professionalization'. When a new speciality arises within a bureaucracy, it normally involves only a few people who can maintain fairly close contact with each other, and whose marginal position provides an incentive for cooperative action. The few specialists are in a position to supervise their successors closely, and if necessary to train them personally. If the demand for employees with the new skill grows rapidly, informal means of communicating values and technical knowledge become inadequate. The formation of a professional association or institute which gives diplomas helps to achieve several goals. If the diploma becomes accepted as the recognized qualification in the field, it gives the institute's leaders—who award themselves honorary fellowships and diplomas—considerable control over the training of their subordinates, both in terms of their technical competence and their acceptance of the values which are regarded as appropriate. It enables the embryo profession to control entry to its own ranks, and possibly to limit this in the interest of creating a shortage and pressing for greater rewards. The diploma becomes a symbol of status within the organization. It takes the judgment of competence out of the hands of non-specialist superiors and places it in the hands of the association or institute; and it can be used as an agent for collective bargaining. The advantages do not accrue solely to the new professionals. The bureaucracy is freed from the responsibility of having to judge technical competence in a new and changing field, because the graduate with a diploma carries a type of guarantee of at least reasonable competence and morality. Furthermore, the association or institute may organize an employment service; it thus gives the bureaucracy access to a list of available competent persons, and may greatly simplify its task in recruitment. Such services may be very valuable to bureaucracies; in earlier times, a corresponding guarantee was often obtained, because education and recruitment was restricted to a very limited section of the population, and only 'gentlemen' would be considered. Wealth, family connections, or attendance at particular schools or clubs may also function as similar guarantees.

The second question arises in many connections: what happens when two or more candidates are equally qualified in terms of the formal requirements for the position? Caplow and McGee[25] explored this problem in some detail. Their work suggested that the choice was often made on 'irrational' criteria: appointing the man who will not 'rock the boat', the man who is recommended by a friend or who went to the more prestigeful school, the man whose wife is more socially acceptable, or the man whose leisure interests and club memberships coincide most closely with those of the present members. The importance of 'irrational' criteria is usually difficult to assess, though it gives rise to many deeply held convictions. Alternatively, the condition of equality may arise because one candidate is superior in one respect, another from a different point of view. The requirements themselves may be somewhat flexible or even inconsistent; this allows the recruiters more freedom in the selection of candidates. Caplow and McGee recognized, as an alternative solution, that the recruiters were likely to be influenced by the prestige which the new member might be expected to bring to the department.

Turning to judgments by co-workers rather than employers, impersonality implies judgment by results, and not by need. Judgment may be applied primarily to the individual, or only to the department as a whole. Although the 'output' of many bureaucracies is not susceptible to ready measurement, attempts are usually made to obtain data which are relevant to the assessment of 'output'. Blau's study of a government employment exchange[26] showed that a major effect of such data-gathering could be to distort the goals of the organization, and to induce dysfunctional behaviour. When the clerks were assessed competitively in terms of the number of placements they made, they tended to hoard notices of vacancies, to pressure applicants into jobs of questionable appropriateness, and to pass over clients who needed intensive help and who presented problems. Although serious distortions may result, the principle is becoming more widely accepted that the bureaucrat should be judged by his performance; and that other criteria such as seniority are relevant only in so far as they can be shown to be productive of valuable experience.

The emphasis on judgment by results has several consequences for

[25] Caplow, T. and McGee, R. J. *The Academic Market Place.* Basic Books, 1958, chap. 8.
[26] *Op. cit.*, pp. 344–348.

the organization. Firstly, it attaches a premium to those departments which can readily demonstrate their results in tangible form, at the expense of those whose results are less tangible: the highway department, for instance, at the expense of the public library. Burton Clark[27] has shown this in relation to the dilemma of the adult education administrator. Courses tend to be strictly tied to be public demand, because those which are not self-supporting are quickly discontinued out of deference to the treasurer's department and the elected representatives. At the same time, the adult education administrator has a low status among educational administrators and is criticized because he puts popularity before educational principles in planning his courses.

Secondly, judgment by results is apt to mean that people or objectives which do not fit readily into the organization's framework are moved around from one department to another, often without any plan for dealing with them. This quickly gives outsiders the impression that the bureaucrat is either cowardly, and wishes to avoid responsibility; or else he is incompetent, because he does not know how to do his job.

Thirdly, the emphasis on judgment by results means that as professionalization develops, the relationship between a bureaucrat and his co-professionals becomes increasingly critical for his chances of promotion. Within the professions the euphemism 'colleague' is widely used to define this crucial relationship.[28] Becker[29] has offered an insightful analysis of the powerful social control which this relationship can impose and the fear that a colleague may use his position as a qualified judge to criticize one's own professional competence. The relationship implies an expectation of willingness to support one's peers uncritically in any public situation where they may be subject to criticism by 'outsiders'. To a much smaller extent, it also implies an expectation that one will defend one's immediate colleagues against criticism by other professionals. It implies also a policy of mutual respect for the professional's independence and judgment, which involves scrupulous non-interference in the colleague's professional work unless one is strongly pressed to offer comments. In Becker's example, it was taboo for a teacher to enter the classroom of another during a lesson, or to question subtly the grade given by another

---

[27] Clark, B. R. 'Organizational Adaptation and Precarious Values'. *Amer. Sociol. Rev.*, vol. 21, 1956, pp. 332–336.

[28] In acknowledgements in professional journals, the term is frequently laden with irony based on fear. The Colleague, like the primitive god, is a potentially dangerous and powerful figure, even if he is a status-equal; one hopes that if he is praised publicly for his stimulating comments, and even wise guidance, that perhaps he will judge one more favourably when the opportunity arises.

[29] *Op. cit.*, pp. 249–250.

teacher. This is not to suggest that professionals regularly fulfil each other's expectations and form a mutual admiration society. They do, however, expect to draw a clear line between legitimate private and illegitimate public criticism; and are apt to become extremely upset when these expectations are not fulfilled. The 'opposite number' can be regarded as a particular form of colleague relationship with a member of another organization or department.

Turning finally to the judgments by the client or applicant for employment in the bureaucracy, the situation is again complex. As Gouldner[30] has pointed out, the reception which the client or job applicant receives is likely to depend on a variety of circumstances: the state of the labour market, the degree of monopoly which the bureaucracy has, changes in opportunities for social mobility, absentee or local ownership and the ratio of overhead to labour costs. To these one might add the class status of the client or applicant, the status of the bureaucrat he meets within the organization, and the client's or applicant's status in the specific situation—as a valued customer, as a poorly qualified applicant for employment, as a welfare beneficiary, as a chief officer in a large local authority.

Further, the impersonality which is proclaimed as the ideal by the organizational rules may conflict with the client's desire for personal attention. The client may well perceive himself as a special case, either in the sense that he does not perceive himself as fitting neatly into an available category, or more radically as a person deserving special attention and favourable treatment. In particular, those clients who spontaneously visit the bureaucracy's office building in person, or who write regularly, are likely to be both the clients with whom the bureaucrat has the most contact, and also those who expect the most personal attention. Such a conflict of expectations is apt to be seen by the bureaucrat as a challenge to the authority of the rules under which he works, and as a threat to his impartiality. The client, by contrast, might well be shocked by the insinuation that his conduct was in any way inappropriate. It may be possible for the bureaucrat to resolve this situation by the exercise of his discretion in applying the rules; and where long-term cooperation and support is desired, he may be anxious to do this. In other instances, he may seek to frustrate the client, if he cannot convince him, to the point where the client abandons his quest, or loses his temper and the argument. It is possible to conceptualize the outcome as a bargaining process, which follows the organization's rules rigidly in only a minority of instances. At the same time, one should not expect a simple relationship between the threat which the client can potentially muster and the special consideration he receives. The large business firm may act as if highly

---

[30] *Op. cit.*, pp. 399–400.

sensitive to the prospect of offending a single customer, or it may gear its response strictly to the client's power to influence its sales.

Again, therefore, the extent to which the expectation of impersonality is attained depends on a variety of factors; and the correspondence between formal expectations and actual behaviour may be weak. As was stressed in the first section of the chapter, the model of a bureacracy as a highly efficient, impersonal machine is thoroughly applicable only under certain limiting conditions. It may apply differently at different levels in an organization; and it is especially likely to prove inefficient when the administrative categories fit poorly, the number of cases per official is limited, and administration requires interaction over a period of years between a particular official and a particular client.

3. Division of labour, with relationships as means to ends; dominance of the large organization; codes of ethics

For Weber, an elaborate division of labour was one of the defining characteristics of a bureaucracy; in pure bureaucracies this was combined with a careful attempt to establish fixed areas of jurisdiction and clear responsibility. The division of labour was seen as a more rational and efficient method of attaining the organization's stated goals; and was assumed to have been set up deliberately for this purpose. The division of labour implies that coordination of diverse activities becomes a critical problem.[31] Although an elaborate division of labour means that a person is dependent on a great number of others, and is dependent on others in many ways, it does not guarantee that these activities will be articulated with each other. More important, it tends to complicate and frequently to obscure the relationship between the participant's activity and the group's goal, and thus to reduce the probability that the participants will readily identify their own interests with those of the organization. Integration presents a critical problem for the organization with an elaborate division of labour; it requires either an elaborate supervisory structure or the skilful harnessing of informal group interests to the service of the whole organization, or if possible both.

While the purpose of the division of labour is the achievement of the stated goals, the relationship between means and ends is unlikely to be as simple in practice as it appears on paper. The means may

---

[31] While the problem becomes critical, according to Parsons (*The Social System*, Free Press, 1951, pp. 507–508), with increasing size, this need not imply that size leads to a greater proportion of employees being engaged in coordination. See the discussion in Mott, P. E. *The Organization of Society*. Prentice-Hall, 1965, pp. 65–66; and the comments of Gouldner, A. W. 'Metaphysical Pathos and the Theory of Bureaucracy'. *Amer. Polit. Sci. Rev.*, vol. 49, 1955.

come to be treated as more important than the goals, and the perpetuation of the organization in an unchanged form may come to occupy more of the bureaucrat's attention than the achievement of the official goals. The means may be relatively inefficient as techniques for reaching the goals, because they may be based on unrealistic assumptions about motivation and behaviour, or because they ignore important human factors. Indeed, the deviant practices of peer-groups may sometimes be more efficient than the officially prescribed practices. The formal division of labour may come to be dysfunctional if it is carried too far and induces restriction of 'output'.

Dreyfuss[32] has taken the position that the main function of the division of labour is as an agent of social control, a technique by which the owners and/or managers can divide and rule. He takes the view that the division of labour has been carried far beyond the technical needs of the situation. Its main effect has been to give the employer and employees alike an added sense of their own prestige and power. Although this is in large measure illusory, he argues, the sense of power and the competition for small promotions encourage identification with the objectives of the superiors. This last point appears rather less convincing than Dreyfuss' earlier arguments. If Dreyfuss is referring to the elongation of the hierarchy as a result of the division of labour, research suggests that the incentive may be effective among those in intermediate statuses, but is unlikely to be effective among those in the lower statuses. If he is referring rather to the broadening of the hierarchy to include more distinct statuses at a given level, the division of labour is likely to obscure the clarity of status differences, but may have little effect on loyalty. Only in so far as the more elaborate division of labour increases the proportion of intermediate statuses is loyalty likely to be affected.

The second part of this proposition sees a close relationship between the growth of bureaucracy and the dominance of large organizations. Several writers have noted that, while there is probably such a relationship, the increasing size of the society is not an adequate explanation of pure bureaucratization. Gouldner, for instance, gives examples of small organizations which are bureaucratically organized, and of large organizations which are not bureaucratically organized.[33] Wilensky's account of the social organization of four American trade unions[34] showed that although pure bureaucracy was spreading, it has not yet become the dominant pattern within these unions. The division of labour was often far from clear; loyal officials' wives were often given significant posts; job definitions were rarely precise; and

[32] Dreyfuss, C. *Occupation and Ideology of the Salaried Employee.* (Eng. trans.), 1938, pp. 17, 75–77.
[33] 'Metaphysical Pathos . . .', p. 498.
[34] *Op. cit.*, pp. 256–258.

frequently the hierarchy of authority could not be represented in a clear organizational chart. Although increasing size creates pressures which favour a different type of organization, growth in size is not a sufficient explanation of bureaucratization. These arguments suggest that, even in urban-industrial societies where pure bureaucracy is common, it is not inevitable. It is still arguable that size leads inevitably to bureaucratization, however, if one includes the patrimonial type as an alternative outcome. If this argument is accepted, of course, it is still necessary to examine the conditions under which one type rather than the other emerges.

While the dominance of large organizations is at first glance obvious, the picture is by no means simple. Although enormous bureaucracies can be found in both business and government, their unity may exist in name only. Authority has frequently been delegated to divisions which have become suprisingly autonomous, and able to resist strenuously any effort to fuse their diverse interests.

The final part of this proposition, the existence of codes of ethics which have an integrative function, is apparent in several respects. Firstly, there is the code of ethics set forth by the organization itself. Sometimes this code of ethics is made explicit: the bureaucrat is expected to be just, dignified, loyal, gentlemanly. In other instances, it is left implicit in the detailed rules which are laid down. This code may vary according to the bureaucrat's status: the higher status person will generally be expected to show greater loyalty, to be more 'responsible'. The lower status person is assumed to be less responsive to such terms, and is accordingly given a more detailed set of regulations; he is expected to be obedient, but not so responsible and not gentlemanly. There may, indeed, be a slightly different code of official ethics at each level in the organization.

Secondly, there is the code of ethics set by the occupational group to which the bureaucrat belongs. For line employees, these are apt to be indistinguishable from the organization's codes, but for staff employees who belong to a professional or trade association, the occupational code of ethics may be quite distinctive. It is likely, for instance, to give top priority to the maintenance of proper professional standards, and to pay little attention to political, administrative or other competing considerations. The distinction between line and staff will be treated in somewhat greater detail toward the end of this chapter; here we are concerned only to draw attention to the existence of a separate set of codes of behaviour.

Thirdly, there is likely to be a set of informal codes of ethics. These are not always readily distinguishable from the occupational codes, unless the occupational groups cut across status lines. The most striking feature of the informal code is its interpretation of loyalty to the informal group. Loyalty in the formal organizational code means the diffusion and conscientious execution of official policy; and loyalty in the professional code is interpreted as the maintenance and diffusion of standards of competence. Such diffusion is expected to take place both within and between status levels. Loyalty to the informal group, however, is interpreted as the restriction *within* a status level of information, especially information about behaviour which deviates from or is irrelevant to the rules. The informal code also includes judgments on the relative weight which should be attached to particular rules, the extent to which they can be ignored with impunity, and the unwritten rules whose observation is important.

While these codes of ethics have integrative functions, it is clear that the professional and informal codes in particular integrate only segments of the bureaucracy; and if they combine loyalty to the in-group with contempt of the out-groups, they are likely to be devisive for the bureaucracy as a whole. Their impact on the rules may be either functional or dysfunctional. Where professional and informal codes tend to support the formal, they will have a strongly integrative function. Indeed, in this event the official rules may become institutionalized and be transformed from means into ends. Myers refers to the instance where the employee attaches much significance to traditional practices that he 'unselfconsciously accepts his own presence on the job as a guarantee of proper . . . practices.'[35] Such a transformation may make the dismissal or resignation of the employees necessary if changes are to be introduced successfully. Where occupational or informal codes contrast with official codes, of course, the conflict is likely to be a source of social change.

4. Division of labour grows with the market; extreme specialization, interdependence, unstable equilibrium

The fourth proposition appears eminently reasonable in urban-industrial societies, though its validity in pre-industrial societies is open to question. Two queries do, however, arise in relation to the first part of the proposition. The first concerns the range over which the relationship holds: given that the market for a product or service may grow in proportion to the population, without obvious limits, is an increase in the division of labour necessarily an efficient response? The second concerns the nature and determinants of the relationship, in those circumstances where it holds: under what conditions does the division of labour grow faster than, slower than, or at the same pace as the growth of the market? Does the nature of the product or service affect the profitability of increasing the division of labour in response to a growing market? Where a non-profit-making service is involved,

---

[35] Myers, R. C. 'Myth and Status Systems in Industry'. *Social Forces*, vol. 26, 1948; reprinted in Merton *et al.*, *op. cit.*, p. 277.

does the relationship still hold without modification? Are there sources of resistance to an increased division of labour, in situations where it would be profitable for the organization?

There is some evidence to suggest that an increasingly elaborate division of labour within a firm does not always increase efficiency. It establishes additional interest-groups within the organization, complicating the problem of coordinating activities and interests. It may reduce the skill and decision-making required for some jobs, to the point where they remain unfilled or engender low morale and low productivity. It presumes, as Merton has pointed out,[36] a disciplined, methodical performance of one's own duties and a keen sense of one's limited authority—a combination of qualities which becomes increasingly crucial as the division of labour becomes more pronounced. Areas of responsibility may not coincide with areas of control. As Gouldner and Dreyfuss have argued,[37] the division of labour may become too elaborate for efficiency: the organization may become artificially complicated. The major function of the division of labour may come to be the avoidance of dependency on particular workers, as their tasks are made more simple and employees cease to be irreplaceable. Where the worker's task is simplified, his bargaining position and indispensability are weakened; whereas the position of his coordinators may be strengthened as their tasks become more complex.

The relative rates of growth of the market and the division of labour have been little studied; only a few remarks will be made here. Firstly, historical studies of particular firms are unlikely to reveal a consistent relationship. Sills' work on nation-wide voluntary organizations[38] underlines the critical significance of certain key decisions for later growth; an organization's division of labour may not be responsive to small changes in the market once it has been established. Secondly, advance planning frequently requires that reorganizations take place in *anticipation* of changes in the size of the market; where predictions are subject to considerable margins of error, the adjustments between an organization and its market are apt to be slow and jerky. Thirdly, the composition and geographical distribution of the market may be more important than its sheer size in influencing the division of labour. Consequently, organizations with markets of similar size may be organized very differently. Fourthly, the type of product or service influences the extent to which division of labour is profitable, for a given size of market: some are readily standardized, and can profitably be divided into their constituent parts while the market is still small. In other cases, a much larger market may be needed

---

[36] 'Bureaucratic Structure and Personality', p. 365.
[37] 'Metaphysical Pathos . . .'; Dreyfuss, *op. cit.*, pp. 75–77.
[38] Sills, D. L. *The Volunteers*. Free Press, 1957, pp. 253–270.

to make an equivalent division of labour profitable. While the proposition may hold in broad terms, then, it requires considerable qualification before it can be applied to all cases.

The second part of this proposition relates the elaborate specialization and interdependence to instability in the bureaucracy. Its validity is likely to depend on the level at which instability is sought. On the one hand, the large and complex firm is unstable, in so far as its operations may be brought to a halt by a small but determined occupational group which begins an unofficial strike. On the other hand, Caplow[39] has shown that large firms are less prone to bankruptcy than small firms. His explanation for this, that the number of ways in which the employees can be reshuffled grows far faster than the number of employees, is neither convincing nor very relevant here. An alternative explanation—that large firms are better able both to hold their market against competitors and to diversify their products—is relevant in so far as diversification involves a more elaborate division of labour. The same argument applies of course to bureaucracies which do not sell their products directly: a more diversified organization can relatively easily expand some sectors and contract others as the situation changes. The stability of the large organization in the face of external changes may be more apparent than real: disproportionate expansion of some departments and the abolition of others is rarely a smooth operation, although it is likely to be much less turbulent if the social relationships among those involved are mostly impersonal. Finally, Selznick[40] has argued that increasing specialization leads to *greater* instability, in so far as it implies greater difficulty in designing an ideal organization. Where people interact as wholes, and not in single role-relationships, non-rational elements are indispensable in the coordination of their activities; yet the complex division of labour challenges these non-rational elements by working on the assumption that they are either unnecessary or capable of spontaneous adjustment to the needs of the organization.

Several writers have examined the relationship of specialization to stability through their mutual links with autonomy. To Weber, bureaucracy implied the specialist's autonomy within his area of competence, and his security against dismissal except for gross incompetence or immorality. Formally, then, the specialist's position is extremely stable. In practice, as we have stressed, informal security is at a much lower level: the specialist's post or department may be abolished; his privileges and freedom may be informally curtailed; he may be pressured to resign; his informal status may be systematically

[39] Caplow, T. *Principles of Organization*. Harcourt-Brace, 1964, p. 34.
[40] Selznick, P. 'Foundations of the Theory of Organization'. *Amer. Sociol. Rev.*, vol. 13, 1948.

reduced. He may thus be placed in a position where he 'sees no alternative' but to resign.

Wilensky[41] challenged Wirth's position radically on this point, although the trade unions he studied were certainly not pure bureaucracies. He argued that stability does not come from exclusive preoccupation with one's specialized sphere of competence: the official who has autonomy (and security, one might add) is the one who can make alliances with other officials outside his field. Since Wilensky was not studying pure bureaucracies, it is unclear how far his argument applies to them; it is, however, much more likely to apply to staff chiefs than to other staff members.

Becker and Levenson[42] have examined interdependence in bureaucracies from the viewpoint of stable relationships between people at different status levels within the same department. Becker drew attention to the existence of an elaborate system of social controls between teacher and school principal, which limited the extent to which each could dominate the other. The work of each could be seriously impeded or assisted, and the reputation of each damaged or improved, by the behaviour of the other. This became most apparent when contrasted with the relationship between teacher and parent, in which there were no corresponding mutual controls, and in which the teachers studied expected the principal to defend them loyally against any complaint by the parent. He commented on teachers' defensiveness in relations with parents:

'The relations of ... functionaries to one another are relations of mutual influence and control, and ... outsiders are systematically prevented from exerting any authority over the institution's operations because they are not involved in this web of control and would literally be uncontrollable, and destructive of the institutional organization, as the functionaries desire it to be preserved.'[43]

This picture is an attempt to express general expectations and fears, of course, rather than to describe reality precisely. Parents' behaviour becomes uncontrollable only where it is expressed through non-institutionalized means, and is therefore less predictable. Where it is expressed through institutionalized channels, a network of mutual obligations is likely to develop quickly, which will offer defences against arbitrariness, even though it does not shield the administrator or the teacher from conflicting political pressures. The significance of unpredictability is two-fold: firstly the teacher is uncertain what

[41] *Op cit.*, pp. 231–235.
[42] *Op. cit.*, pp. 246–249; Levenson, B. 'Bureaucratic Succession', in Etzioni *op. cit.*
[43] *Op. cit.*, p. 251.

steps the parent will take in support of his view, and what steps need to be taken to defend her action; and secondly, the teacher is likely to be more uncertain how far the parent's complaint has the support of other parents who may take organized action against her.

Levenson[42] examined interdependence between the levels in terms of sponsorship and promotion. Extreme specialization raises the problem of training a limited number of successors for each specialized post in advance. He argued that promotion in the bureaucracy is liable to depend on 'sponsorship' in cases where several candidates meet the formal criteria. Sponsorship is clearly valuable for the subordinate, in that it increases his chances of promotion in cases of genuine doubt and in cases where the office-holder is asked to advise on the choice of a successor. It is also valuable to the superior, who is able to claim loyalty in return for his sponsorship—both personal loyalty and assistance in ensuring that his requests are carried out and his reputation enhanced. Where sponsorship is significant, a bureaucrat's promotion prospects are likely to be influenced, not only by his merits, but also by the 'promotability' of his immediate superiors.

5. Impossible to assemble all employees together; indirect communication by mass media and special interest groups

While the mass media may be a significant source of information about other parts of one's own organization, printed material 'for internal consumption' is probably more significant. Among the special interest groups which influence the flow of communication in a bureaucracy, the professional and trade associations to which its employees belong are perhaps particularly important. Since this fifth proposition is not difficult to accept, it will be more instructive here to examine some of the processes of communication in bureaucracies, and some of the groups which influence communication in particular directions. Direct communication to all members of the organization is unusual, and normally reserved for ceremonial occasions, where the setting is formal and the communication only 'one-way'.

The rules of the organization represent the most obvious form of indirect communication: they are transmitted downwards in a fairly permanent form, and are impersonal in the sense of being addressed to office-holders rather than to individuals. In return, formal reports are transmitted upwards. The process is frequently slow, because the communication must normally pass through each intervening level in the hierarchy, and may be referred back at any stage for clarification. Since there is often a taboo on threatening the authority of an office-holder by by-passing him, the report or rule is apt to accumulate dust and revisions as it moves upward and downward towards its ultimate destination. The most conspicuous exception to this leisurely pace is found when a request is made by an elected official. There may be a short time limit within which the department is expected to prepare a reply, and the answer may be embarrassing or the data inaccessible. My own impression, from participant observation in a fairly large local government department, was that such requests struck the department like thunderbolts. Word of their contents used to travel quickly to the lowest levels; and all routine activity ceased until the request had been carefully examined. If it was neither embarrassing nor complex one heaved a sigh of relief that the crisis had passed; otherwise all hands were mustered in an emergency operation until the information or a suitably worded answer had been assembled.

The existence of an elaborate hierarchy generally implies that communication will be indirect; while this permits fuller explanation and closer supervision of instructions from superiors, its sources of inefficiency should not be overlooked. Barnard[18] pointed out that indirect communication was more conducive to the issuance of conflicting instructions, and to instructions which the recipient perceived to be contrary to the organization's goals. It can also be inefficient in that it introduces an additional barrier to the flow of communication. As peer-groups develop at every level, communication is most seriously restricted in organizations which have an elaborate hierarchy; and superiors may be systematically deprived of the information which they need to make decisions. The intermediate official may 'face both ways', creating ignorance and misunderstanding between those above and those below him; he may thus become a kind of special interest group. This draws attention to the phenomenon of 'bureaucratic sabotage',[44] whereby the content of a communication is not only transformed but reversed as it passes from source to executor, yet word of this change normally does not penetrate back to the source unless a scandal develops. 'Sabotage' may occur equally to communications going up or down the hierarchy; it is not the prerogative of officials at any one level.

The role of the peer-group in relation to bureaucratic sabotage deserves further study. The peer-group is perhaps unlikely to cause sabotage directly: both are probably the result of the same situational pressures. The peer-group may, however, be crucial in organizing sabotage and in providing a legitimation for it. Such a role would be consistent with the peer-group's function in limiting the flow of communication upwards and downwards, and in restraining discussion by the public or in the presence of subordinates of the means by which decisions are made. Barnard[18] and Gouldner[45] have pointed out some of the circumstances in which bureaucratic sabotage is likely, and have isolated some of the situational pressures which may be

[44] Brecht, A. 'Bureaucratic Sabotage'. *Ann. Amer. Acad. Polit. Soc. Sci.*, Jan. 1937.
[45] Gouldner, A. W. (ed.) *Studies in Leadership*. Harper, 1950, pp. 644–659.

responsible. Barnard argued that the acceptance of instructions was most certain when seven conditions were met. The instructions should: (a) be understood; (b) be believed by the recipient to be consistent with the organization's purposes; (c) be believed by the recipient to be compatible with his own interests; (d) be believed by the recipient to be within his capability; (e) fall within the range of instructions which the recipient accepts without question as legitimate requests; (f) be traceable to a source which the recipient regards as authentic; (g) be authoritative, in the sense of emanating from a person who is either technically competent in the area of the communication, or is formally empowered to make it, or both. One may reasonably infer that resistance to the execution of the instruction will grow as these conditions are broken. One should be careful not to infer, of course, that compliance rests mainly on the whim and self-interest of the subordinate; nevertheless Barnard's account is instructive in illustrating some of the sources of sabotage, and in stressing that compliance is by no means automatic. Gouldner suggests a different source: intentional sabotage, at least, is most likely when a new superior is unaware of the informal understandings and relations which already exist among his subordinates; and when he is faced with a heritage of informal expectations and promises which his predecessor was unable to fulfil.[46]

Although we have stressed here the role of the informal group in altering communication, this should not leave the impression that peer-groups always operate to the detriment of the organization. In some circumstances they may have important supervisory functions in enforcing the rules; in others, the changes they initiate may be more efficient than strict adherence to the rules, in achieving the official goals. 'Sabotage' is not therefore the unavoidable result of peer-group development; nor need it inevitably await the development of peer-groups. Where such groups do form, however, they add both organizational skill and legitimation to deviant behaviour; and their spread increases both the range of channels through which communication flows and the possibility that a bureaucrat will be exposed to conflicting communications.

6. Differentiation and specialization increase as density grows
Wirth's sixth proposition again defies assessment under controlled conditions: the growth of bureaucracies is normally a complex matter, in which the isolation of simple relationships is difficult. One or two observations can, however, be made, about the possible effects of an increase in density with no change in the size of the total population. Firstly, the increase in density reduces the possibility that two per-

---

[46] *Ibid.*, p. 651.

sons will be carrying out the same task unknown to each other. Increased density is likely to augment both unavoidable communication and friction, and consequent awareness of others' activities. Specialization represents one method of reducing these frictions, by giving each rival a distinct sub-area within which to exercise autonomy. Secondly, the organization may respond by approximating more closely the pattern of a total institution. Where this occurs, specialization is likely to decrease, but differentiation between statuses becomes more pronounced. The total institution, as Goffman has argued,[47] is an economical way of managing the movements of a large number of people in a small space with limited resources.

7. Close physical but weak social bonds; readily perceptible symbols
The formal organization and physical layout of a bureaucracy is generally based on the assumption that close physical proximity is conducive to good communication, while weak social bonds are conducive to competent and impartial administration. This assumption needs substantial qualification before it can be accepted. Firstly, close proximity implies unavoidable interaction, which makes the maintenance of status differences more difficult. Secondly, the maintenance of social distance increases the barriers to the communication on which decision-making depends, and which physical proximity was designed to minimize. In both these respects, then, physical proximity and weak social bonds have opposite effects and are liable to cancel each other out.

Five solutions to this dilemma appear to be possible. Firstly, one may try to maintain both close physical contact and impartiality, and to combine the two as judiciously as circumstances permit. Secondly, one may specifically aim to maintain both by the use of elaborate ritualism in all interaction. Thirdly, a double leadership structure may form, in which the chief remains the socially distant task leader, while the deputy chief becomes the physically close socio-emotional leader. Fourthly, the organization may stress formality and distance, at the risk of poor communication, but without developing elaborate rituals. Finally, it may stress the strengthening of social bonds, at the risk of undermining formal authority. It is perhaps worth noting briefly that close physical and weak social bonds also characterize mobs and crowds; only the formal organization, which sets up common goals and creates informal interest groups within its framework, distinguishes the bureaucracy from the mob or crowd. Indeed, mob activity is sometimes likely in bureaucracies such as custodial prisons which stress this combination of close physical and weak social bonds, if they also have few agreed formal goals.

[47] *Op. cit.*, pp. 312–316.

We have already seen that weak common bonds are by no means the rule, especially among people at the same status level; and we have looked at some of the factors which influence the formation of stronger bonds. Here it is perhaps necessary only to discuss briefly the complex relationship between physical proximity and the development of informal social ties. This relationship can perhaps be summarized in two propositions, though these need much more exploration and precision than can be given at present. Firstly, informal ties tend to form to smooth all unavoidable contacts; but once formed, they tend to develop goals of their own which may become more significant than the source of the original contact. Secondly, the development of social ties is not automatic, but depends on a set of supporting conditions. The first proposition can be illustrated by sociometric studies of initial interaction in the office or the student dormitory. Newcomb,[48] for instance, found that initial interaction and friendship choices were very strongly influenced by physical distance; only later did interaction and friendship come to span greater distances. Similarly, most studies of housing estates have stressed the great importance of proximity in affecting interaction. In the latter context, it has sometimes been possible to relate this pattern of choice explicitly to fear of inharmonious relations with those who were nearest to the respondent. Morris and Mogey[49] explored some of the effects on cohesion and neighbour relations when interaction did not pass beyond this initial stage.

The two most obvious supporting conditions for the development of social ties are the existence of common interests which are threatened and the absence of clear status differences. Most studies of output restriction have linked this explicitly with the fear of redundancy or a cut in piece-rates. Similarly, social ties frequently form between bureaucrats at the same status level in different departments, if their work requires consultation and collaboration. The significance of status differences was apparent in Blau's work.[50] Cooperation among clerks employed in the same office on the same tasks tended to spread only among those with approximately equal skill and experience. The newer members tended to seek advice, not from the most experienced clerks on the same status level, but from others with limited experience; in this way it was possible to develop a reciprocal relationship, whereas the experienced clerk was rarely in a position to require the services of the inexperienced.

The second part of Wirth's proposition refers to the use of readily

---

[48] Newcomb, T. M. 'The Study of Consensus', in Merton, Broom and Cottrell, *op. cit.*

[49] Morris, R. N. and Mogey, J. M. *The Sociology of Housing*. Routledge and Kegan Paul, 1965, ch. XII.

[50] Blau, *op. cit.*, pp. 346–347.

perceptible symbols as a method of maintaining social distance and (to a smaller extent) social differentiation in general. It has already been suggested that an elaborate system of symbols and sanctioned ritual behaviour may maintain status differences among those in close contact for a considerable period. Most analyses of bureaucratic symbols have focused on those which separate different status levels; but one should not overlook those which differentiate among those at the same status level. The symbols may be of many kinds, and need not be limited to what is readily visible, as Wirth had originally written. The uniform, the diploma, the wall-to-wall carpet, the executive desk and the clean-shaven appearance are important; but so are the use of language, accent, the tone and the loudness of the voice. Appeals to the sense of smell may also be apparent, through the use of cosmetics, scented soap, mouthwashes and deodorants. Additional symbolic representations of status are found at *rites de passage* and other ceremonials which have been carefully analysed by Caplow.[51]

These symbols and rituals may be functional not only in regulating unavoidable contacts and maintaining status differences, but also in simplifying communication between members of different groups. Status symbolism, for example, may be useful in judging whether a communication is authentic, whether it is authoritative, and whether it will be intelligible to the recipient. It can clearly become dysfunctional, of course, when symbolic and ritual behaviour is interpreted as reflecting actual beliefs. Stouffer found, from a number of indicators,[52] that officers, who were accustomed to ritual respect and obedience from subordinates, were inaccurate in predicting the subordinate's opinions. The chief who is not watchful may receive only carefully guided tours of his own department, and reports which indicate what he would like to believe, rather than what actually happens.

The same phenomenon can be observed in the college classroom: the students sit quietly, interrupt rarely, and frequently give the appearance of being busy note-takers. The more interested and responsive students tend to sit near the front, to smile at the appropriate moments, to ask questions when expected, and to visit the lecturer spontaneously outside class hours to express interest. Those who are bored tend to stay away or to sit near the back, where they are less observable. These rituals may be functional in reinforcing the authority and status of the lecturer, especially if he is nervous, and in encouraging him to continue to perform his role. They may, however, be dysfunctional in giving him the illusion that his class is all keenly attentive and well able to understand the material he presents. In the

[51] *Op. cit.*, pp. 85–87.
[52] Stouffer, S. A. *et al.*, 'Barriers to Understanding between Officers and Enlisted Men', in Merton *et al.*, *op. cit.*, pp. 267–269.

latter case, the exams bring their round of disillusionment; but it may then be too late.

Finally, the importance of symbols requires procedures for distributing and controlling access to those which the bureaucracy itself can award. The procedures for making and applying for awards may be formal, and symbols may be associated specifically with one status level. In most cases, however, the maintenance of the status distinctions between one's peers and one's subordinates rests with the peer-group. The peer-group attempts to retain control over access to certain types of symbol; and where the symbols can be purchased outside the organization, to discourage their use among subordinates or to disvalue them altogether. Discouragement can sometimes be backed up by informal sanctions against the subordinate who claims symbols to which he is 'not entitled'. Where the symbols are owned by the organization, an official may be assigned to proctor their distribution. Turner[53] has given an excellent account of the conflicts between formal and informal expectations in the behaviour of the officer who guards the organization's status symbols.

The function of the peer-group with regard to status distinctions between one's peers and one's superiors is perhaps more complex. Again the peer-group encourages conformity among those at the same level, but its attitude towards the acquisition of the symbols of higher status is sometimes ambivalent. On the one hand, there is frequently a strong tradition against the member who aspires to the symbols of higher-status groups single-handed, and a variety of pejorative terms have been invented to describe the social climber. On the other hand, collaborative efforts to acquire at least some of the status symbols of the higher groups are likely to be encouraged. In consequence, the more readily accessible symbols frequently descend the formal scale fairly quickly, while those which are expensive or difficult to reproduce retain their significance for longer periods.

8. Economic competition determines land use; residential desirability is complex; segregation of commercial and residential land uses

We mentioned earlier that relatively little work has been done on the ecology of bureaucracies, although the concepts of ecology are clearly applicable. Although few bureaucracies offer residential facilities, the desirability of sites for office buildings may be both influential and complex; similarly, one may observe physical segregation, not only of departments, but also of different land uses—factories, warehouses, head and branch office buildings—which belong to the same organization. While the usefulness of the concepts is clear, the absence of data means that one can only offer tentative observations.

[53] Turner, R. H. 'The Navy Disbursing Officer as a Bureaucrat'. *Amer. Sociol. Rev.*, vol. 12, 1947.

The importance of economic competition in determining land use is likely to be highly variable. The status of a department within a bureaucracy, and its ability to gain facilities and space, may depend on its income-generating power. Status is influenced by many other factors, however; and the relationship between status and ability to attain one's goals is by no means simple. Barnard[54] suggested that high status accrues to those who show exceptional ability in solving exceptionally difficult but exceptionally important problems. Caplow[55] modified this with the view that all organizations have 'points of stress'—values or basic group needs which present especially intractable problems. In a maximum security prison, for example, the problems of keeping the inmates clothed, fed and healthy, and of motivating them to do the available work, present few difficulties; whereas the problem of maintaining authority over the prisoners, without compromising other goals, is acute. High status accordingly is reserved for the successful security officer, and low status for the rehabilitative workers. The status difference is reversed in prisons which stress rehabilitation and aim to keep authority to the minimum. Although economic resources are important, departments compete also for non-economic resources; their competition may not be primarily through economic means but in terms of values; and in any case their use of economic means is often modified by non-economic values. It is easy, of course, to note limitations to the importance of economic competition; but it would be a mistake to suggest an alternative single type of competition. Caplow and McGee,[56] for instance, come close at times to arguing that competition among American universities can be summarized as competition for prestige; one might draw the conclusion that in this instance, prestige competition determines land use. This is a plausible hypothesis; but to test it one requires an independent measure of prestige, to avoid the circular argument of defining prestige as what universities strive for, and then finding that indeed they do strive for it.

The factors influencing the siting of buildings, especially office buildings, are also largely unexplored. One suspects that in most cases the limitations of the site itself are insignificant. It is important to have reasonable access to major transport routes, and middle class residential areas, and to have good water, sewerage and power supplies; but these restrictions eliminate only a minority of sites. There is much more scope for the type of factors which influence residential desirability; and one would therefore expect two solutions where land is plentiful. Firstly, the prestige office building which is very tall, is

[54] Barnard, C. I. 'The Functions of Status Systems', in Merton *et al.*, *op. cit.* pp. 244–245.
[55] *Op. cit.*, pp. 126–133.
[56] *Op. cit.*, pp. 104–116, 129–138, 145–146.

modern in amenities, but not too modern architecturally; it is located in the main business area, surrounded by other similar buildings and luxury flats, which form a protective barrier between the office and the slum areas which are so often found on the fringes of the central area. If a central site is not necessary for rapid communication with other key bureaucracies, there is much more scope, but perhaps not much more diversity. The second popular alternative is the office in a natural but formally landscaped setting, on a slight hill, modern in its interior but again fairly conservative outside, and not very tall. It is carefully segregated from the factory or so designed that smoke and effluents are concealed, and the office workers have if possible an unspoiled view of nature and the parking lot. The exterior and interior designs both reflect the image which the organization wishes to present to outsiders. In the case of the business firms, this is likely to be an image of cleanliness (implying honesty), prosperity, and conservatism (implying safety for the investors). The image of government buildings in Western countries tends to stress economy-mindedness rather than prosperity; although this is gradually changing. In developing countries, government buildings are apt to be much more striking architecturally, and to convey (at least from the outside) all the aspirations of a vigorous nation making a clean break with the past. Social factors also influence interior design, of course, though this is not central to our argument. One need only note the strategic location of the chief's office, accessible only through his secretary's, which may in turn be enclosed in a small departmental office, typically towards the end of a long corridor. With a few exceptions—notably schools—the physical inaccessibility of the chief corresponds to his insulation from interaction with the man in the street.

9. *Competition and mutual exploitation rather than cooperation at work; life at a fast pace; orderly meaningful routines*
Orderly, meaningful routines are prescribed by the rules of the organization, and enhanced by the unwritten rules of the informal groups; and these have already been considered. The earlier section on communication also outlined the circumstances under which the routines are likely to be meaningful to those who follow, as well as to those who initiate them. The extent to which bureaucrats live at a fast pace is largely unexplored. The stereotyped image of the bureaucrat certainly does not evoke the idea of a strenuous existence, except perhaps when elected representatives raise irate questions. As was mentioned in the third chapter, the impression of working under pressure may be partly a product of unfamiliarity with the work. Familiarity may increase productivity, though this is by no means automatic; but more important, it enables one to predict more easily what constitutes a reasonable day's work, and to judge whether extra assignments are

reasonable. The impact of pressure tends to vary according to one's level in the bureaucracy: for the employee at the lower levels, it tends to mean greater output of the same good or service by the same method. For the higher status employee, it is likely to involve a greater variety of tasks and decisions. Again, there is little research on this topic; though it is self-evident that the increasing government intervention and the greater complexity of contemporary administrative obligations has presented many new forms of strain, and much longer hours, for the higher officials, than in earlier times.

The first part of Wirth's proposition will engage more of our attention; it relates to work relationships, both between and within levels. Some of the factors affecting cooperation and the development of peer-groups have already been outlined; these need not be repeated here. The proposition will, rather, be examined from two points of view: the extent to which the organizational rules presume that competition and cooperation will occur; and the extent to which competition exists within the peer-group.

The rules assume a hierarchy of authority within which each department has well-known areas of responsibility. Normally the rules presume that departments will cooperate, but rarely oblige them to do so. In some instances, specific procedures may be given for settling anticipated conflicts. This may even extend to include conflicts between an employee and his superior.[57] Traditionally, however, these have been regarded as soluble without the intervention of a third party, through the superior's interpretation or clarification of the rules. The ideal is thus a set of rules which covers all conceivable situations. Furthermore, the bureaucratic rules pre-suppose that the rewards of the bureaucrat suffice to motivate him to follow the rules conscientiously, and to refer upwards decisions which are outside his jurisdiction. Consequently they may provide few explicit sanctions against non-conformity.

Organizations vary greatly in the extent to which their rules explicitly encourage or discourage cooperation. If one retains Weber's definition, the pure bureaucracy presumes impersonal cooperation in all cases. To encourage this, it stresses secure and relatively unalterable promotion patterns, at the expense of high salaries. Salary scales are published and annual increments fixed; promotion may be almost automatic. Davis[58] has argued that this pattern encourages conservatism, conformity and lack of enterprise: the official has nothing to gain by an outstanding performance, but may lose if he commits an error of judgment. He is therefore rewarded, in effect, for avoiding

[57] Scott, W. G. *The Management of Conflict. Appeals Systems in Organizations.* Irwin-Dorsey, 1965, p. 6.
[58] Davis, A. K. 'Bureaucratic Patterns in the Navy Officer Corps'. *Social Forces*, vol. 27, 1948.

initiative and passing difficult decisions up to his superiors. At the same time, the effects of rigid promotion should not be exaggerated. All organizations are likely to leave some rewards—by intention or by default—in the hands of the immediate superior, and competition for these remains. Where some assignments are more pleasant than others, there will be competition for these assignments, whether or not there are objective criteria for awarding these. Further, even the most objective criteria available may conceal, rather than removing, the authority of the superior to reward and punish. Finally, competition is inherent in the division of labour, since it inevitably sets up different interest-groups, increasing the likelihood that any particular decision will affect some groups favourably and other unfavourably.

Even when such a rigid promotional pattern exists at the lower levels, advancement among higher-status bureaucrats is not tightly regulated. In this respect, the higher-status official is in the same position as the factory worker or the manager: the rules implicitly stress merit, but give little indication of the criteria by which it is measured. Clearly this encourages much more competition among peers who desire promotion, and increases the insecurity which surrounds it. How far such competition serves the goals of the organization is an open question.

Greater effort is likely to be related to the rewards which it is perceived to bring; but several variables are likely to intervene, influencing the importance of promotion according to merit. Firstly, merit is not always easy to define; and the relative importance of its various facets often cannot be weighed reliably. Hence a poor choice of criteria of merit is likely to distort the subordinates' energies. Secondly, the subordinates' perception of influences on promotion may be seriously inaccurate; yet this perception of influences may be very influential. Thirdly, unrestrained competition is unstable, and where people compete in terms of known, measurable criteria a status order and norms tend to form and to be widely accepted. Once formed, the status order limits the enthusiasm of the deviants for further competition. Competition itself is therefore apt to become conventionalized, and its results highly predictable. The convention may limit competition to certain prescribed situations, and to a few methods. It may also promote the effort to behave in a friendly and informal manner towards one's competitors, especially where competition is most severe and the outcome least certain. Friendly informality is functional, not only in lessening the bitterness which competition may otherwise engender; but also in making cooperation easier at a later date. Where promotions are flexible, the person who is now one's subordinate may later become one's equal or even one's superior. In the same manner, some of the formal courtesies extended by politicians to members of the other party are influenced by the real possibility that those who compete today may need to cooperate tomorrow;[59] and that the party which is larger and in power now may be the smaller, opposition party after the next election.

Sponsorship may fit into this framework in two ways. Where the measurement of success depends in part on the judgment of the superior, rather than on objective tests, there is likely to be competition among those who wish to be sponsored. Where the application of the rules is at the discretion of the superior, sponsorship may imply not only a recommendation for promotion, but may also involve the liberal interpretation and waiving of regulations in favour of the sponsored. Conversely, the superior may interpret the rules very literally or raise additional barriers against the subordinate whom he rejects.

10. Simple class distinctions break down as the division of labour becomes complex

This proposition needs little elaboration. As a bureaucracy becomes more complex, its sub-divisions become more clearly distinguished, and interaction between them is weakened by physical separation and a clearer division of responsibilities. One can readily see the artificial elements in a salary structure which assesses the worth of employees with very divergent skills and qualifications. The imposition of a formal status-structure may create the impression of unity, but it is unlikely to be stable. The rationale for equating, say, the chaplain and the captain is at best very unconvincing, even if it should happen to be based on the salaries which the men could earn in civilian life. An artificially simple status structure of this type is likely to be unstable for two reasons. Firstly, different departments will normally have different growth-rates in response to changing conditions, and this will create pressure to increase salaries more in those departments where new recruits are scarcest. Similarly, the salaries for comparable civilian posts are unlikely to increase at identical rates; and unequal raises may be needed to fill existing vacancies. Secondly, since the rationale by which the chaplain and the captain are equated is necessarily questionable, each occupational group can present a convincing case for more favourable treatment, stressing different criteria of usefulness to the organization.

As departments become more autonomous, their status structures may diverge further, making comparisons more difficult. Departmental autonomy sometimes reaches a point at which coordination becomes exceptionally difficult. The firmness with which local authority housing departments retain control over their waiting-lists is an

[59] Stacey, M. *Tradition and Change.* O.U.P., 1960.

example.[60] Most local authority housing departments resist stoutly the efforts of members of the public or special interest groups to obtain special treatment, as one would expect. They sometimes resist equally firmly the efforts of other local authorities, such as county councils which request housing for teachers. Until recently, at least, most have fought vigorously against efforts to coordinate the waiting lists of neighbouring authorities. Consequently some people who work in one local authority area and live in another are able to apply for council housing in more than one area; others in different jurisdictions are not. Similarly, with very few exceptions, if a person moves from one local authority area to another just before he becomes eligible for a council house, he cannot transfer any of the 'credit' he has already built up. Indeed, he may have to live in the new area for a qualifying period before he is eligible to be placed at the foot of the new list; and at times the qualifying period has been as high as ten years. One is tempted to conclude that this shows unreasonableness and petty jealousy among housing officials. While this may be an element, it is only part of the explanation. After a long period of local autonomy, a great variety of practices have grown up, many of which are now accepted as both legitimate and morally right by their followers. The problem of standardizing them is likely to require either very protracted negotiations or an imposed solution which runs the risk of provoking the resignations of many officials.

Complexity makes the organization more flexible in some respects. As Levenson has argued convincingly,[61] the problem of the 'unpromotable' bureaucrat who is not approaching retirement is thorny, both for his superiors and for his subordinates whose own promotions are somewhat blocked. Since the demotion or dismissal of an employee is likely to affect the morale of many others, there may be no satisfactory solution in an organization with a single status structure. In a more complex bureaucracy, especially one in which considerable turnover is expected, there is much more scope for moving the bureaucrat into a less influential but possibly more prestigeful position. Goldner[62] has analysed such movements in some detail: in the case of a nation-wide organization with branch offices, it may be quite unclear to the bureaucrat and his subordinates whether he has been moved upwards, downwards or sideways. It is particularly likely that one can find a move which will represent a promotion in some respects and a demotion or sideways movement in others. The complexity of the organization increases flexibility in this respect.

[60] Morris and Mogey, *op. cit.*, chap. I.
[61] *Op. cit.*, pp. 367–371.
[62] Goldner, F. H. 'Demotion in Industrial Management'. *Amer. Sociol. Rev.*, vol. 30, 1965.

11. Group loyalties conflict; geographical and social mobility; sophistication
Although loyalty is perhaps indispensable to coordination, secondary groups such as bureaucracies do not normally place such stress on it as do primary groups. Secondary groups are characterized by an individualistic orientation: the individual is *expected* to place his own interest before those of the organization in deciding whether to remain in its employ. Such a course is generally regarded, not merely as rational, but as perfectly legitimate and even preferable. Loyalty to the organization is therefore likely to fluctuate with the rewards which it can offer, relative to other organizations. Within the organization, complexity divides loyalties between the organization, the department, one's profession and its practitioners, and one's immediate peers. Peer-groups will not be discussed further here; we shall concentrate on the distinction between line and staff within bureaucracies, and the conflicts of loyalty to which it is related.

The line-staff distinction can be formulated in a number of ways; perhaps the most useful for present purposes is to equate it with areas of competence. Since this formulation applies to *skills* and not to *positions* in a bureaucracy, and since bureaucratic administration is gradually changing, it will not be possible to describe certain statuses as entirely staff, and others as entirely line. We are concerned with an orientation and not with a position in the hierarchy;[63] clearly the holders of a position may have quite different orientations. Essentially the staff member is the 'cosmopolitan': competent in a particular field, whether this be accounting, chemistry or welfare administration; if his field is sought by a variety of industries and organizations, he can fairly readily transfer it from one to another. His career is not tied to a single organization, much less to a single department. Especially where his field has a professional or trade association, he looks primarily to his professional colleagues for judgments of his competence, and only secondarily to the organization in which he happens to be currently employed. Where his field has become scientific in its orientation, he is likely to value and respond to progress in his field, rather than to the authority based on the traditions of the organization. Since his skills are transferable, his loyalty to the particular organization is unlikely to be firm.

The line member, by contrast, is skilled within an organization rather than a field. His career is tied to a single organization, or a very narrow range of organizations with very similar characteristics.

---

[63] In this respect, we differ from Dalton, though his analysis has been used here: Dalton, M. 'Conflict between Staff and Line Managerial Officers'. *Amer. Sociol. Rev.*, vol. 15, 1950.

He is a 'local', and his power lies in intimate knowledge of its rules and of the relations between its formal and informal structures. Barnard has claimed that such 'local' knowledge is indispensable in running a bureaucracy, and may not even be easily transferable to another large branch of the same firm. This knowledge is based on cumulated experience rather than professional training; and its possessor may not be able to formulate and pass it on easily. The line member has no professional association, though he may belong to an informal 'old-timers' club within the organization. His skill is acquired through long experience rather than academic training; it is more acutely threatened by changes in the structure or in the personnel. The line member is more likely to be on the defensive against the staff member; and is apt to treat the latter as a person on trial, who needs to prove himself and gain informal acceptance within a web of unofficial social control.

As we have emphasized, few employees have only staff or only line characteristics; one can nevertheless analyse the conflicts which arise between predominantly staff and predominantly line members. Dalton[62] identified three conditions as basic to staff-line struggles. Firstly, the ambitious and 'individualistic' behaviour of some staff officers, who were more interested in personal distinction among their professional colleagues than in collaboration with line officers. Secondly, he noted the reluctance of the line officers to accept the staff on the staff's terms, as experts who deserved deference within their own area; and their tendency to assign the staff members (perhaps unintentionally) to tasks where their professional training was of limited use, and where they were less able to distinguish themselves. Finally, he noted the dependence of the staff on line members for promotion to the highest positions, where control over the conditions of the staff's work was exercised. The reluctance of the staff to defer to the judgment of line officials on matters which affected the staff's potential performance and sense of values was more frequent than direct 'interference' by line officials on purely technical matters.

One would expect mobility in bureaucracies to be especially low among line and much higher among staff members, especially those at the lowest levels whose skills have been simplified by automation and the elaborate division of labour. Here we shall be concerned with two consequences of mobility, rather than with establishing its frequency in bureaucracies. Firstly, Gouldner[45] has argued that an organization which has high rates of turnover is likely to respond by organizing more bureaucratically. The rapid changes reduce the time available for training new recruits fully, and weaken informal networks of control. Bureaucracy thus functions as a method of simplifying tasks and of establishing a set of expectations quickly. Secondly, mobility (and more generally communication) between organizations of the same type develops what Caplow calls organizational sets.[64] An organizational set contains a number of organizations of the same type, which are in sufficiently close contact for a status order to develop. An organizational set is less complex than a social system: its constituent organizations share a much narrower range of common goals, and therefore interaction results in a single prestige order rather than an elaborate division of labour. Caplow does not attribute many other properties to organizational sets, though it may be inferred that they function to confine competition between the set members within a few channels.

The third part of Wirth's proposition describes sophistication as a result of conflicting loyalties. This may be illustrated briefly through the process of socialization into and out of bureaucratic roles. One can perhaps isolate for analysis four stages in the learning and unlearning of a new bureaucratic role: these may be referred to as the freshman, sophomore, upperclassman and graduating stages. They arise partly from the necessity of learning and then leaving a role; and partly from the conflict of loyalties, illustrated here in terms of the differences between the formal and informal structures. These stages are not necessarily distinct in time; they are not of specified length; and they are not inevitable—a person may remain fixated at one stage, although one would normally expect him to move through them in order. Finally, there is no implication that each stage represents greater 'maturity' than preceding stages.

The term freshman conjures up stereotyped pictures of utter innocence and this is indeed one characteristic of the freshman stage. The freshman is new to the organization, and is engaged in learning its formal structure and official goals. These present a challenge, and he is proud to internalize them uncritically at first, for evaluation later. At this stage one is proud to know some aspects of the formal structure, such as the names of the chief and his deputy and the formal goals of the department; and learns them indiscriminately. One is at most dimly aware of the informal structure: when it protrudes, it appears as deviant and deserving of reproach. One is disappointed on finding hypocrites and slackers, of whom the formal structure gave no hint, and who one feels should be condemned.

The sophomore stage is characterized by cynicism and overemphasis on the informal structure and deviant goals. These two characteristics are associated: having learned the formal structure thoroughly, one is now learning that in many respects the organization does not operate as it was intended to. Since awareness of the informal structure and its goals are new and incomplete, it too tends to be internalized uncritically at first, and its importance is exaggerated. The

---

[64] Caplow, *op. cit.*, pp. 201–228.

sophomore is proud to tell the freshman 'the truth' about the organization, and to present the view that the formal structure is totally misleading: the bureaucrats are 'really' all money-seeking idle hypocrites, who judge by standards which mock the official ones.

The upperclassman stage comes as one learns to make a more sophisticated appraisal: one has discovered and experienced both structures and sets of values, and can make a more balanced judgment of their strengths and weaknesses. One has developed enough loyalty to the formal organization to have a restrained sense of satisfaction with its achievements and one's own status; and enough loyalty to the informal organization to have adopted the view that its secrets ought to be treated with discretion.

The fourth or graduating stage is sometimes described as desocialization. This stage represents an adjustment to a forthcoming change of status, a rationalization of the decision to leave by rejecting the organization. Especially when the move is voluntary, rationalization of the impending change as thoroughly desirable involves a new emphasis on the unpleasant aspects of the old status; and an innocent, perhaps nervous, exaggeration of the pleasant aspects of the new. One becomes acutely aware of the problems about the present status which were formerly suppressed, and one may wonder how one was able to tolerate the old situation for so long.[65]

The development of sophistication can thus be illustrated as a growing awareness of conflicting loyalties, the simultaneous adjustment to more than one group, and flexibility in suppressing or emphasizing favourable and unfavourable aspects of a role. While the progression through these four stages is neat in the abstract, the transition from one stage to another in practice may be neither smooth nor clear. Some bureaucrats never reach the end of the progression: their experiences in, say, the sophomore stage may discourage them from making a more sophisticated assessment and passing into the upperclassman stage.

12. Levelling influence of mass production; pecuniary nexus

Since this chapter has focused on work relationships, we examine next the possibility that the introduction of mass production and automatic machinery has a levelling effect. If it reduces all manual jobs to very simple routines, this effect is incontestable. In general, however, the effects of automation on status are more complex and little studied. As a framework, one might begin by adapting Caplow's

---

[65] My colleague S. Frederick Seymour has pointed out that this represents a temporary lapse from sophistication, and assists in creating unrealistic expectations for the new role. In this way, the graduation from the old role overlaps with the freshman stage of learning the new role.

propositions on two-person interaction[66] to the interaction between man and machine. Since there is little evidence on their validity, they will be presented with only minimal comment:

(a) Men tend to like the machines with which they interact frequently. Men acquire affection for a lorry, aeroplane or other machine with which they work regularly.

(b) The more frequently man and machine interact with each other, the stronger the man's sentiments of affection are likely to be.

(c) A man who feels a liking for his machine will express these sentiments over and above the activities required by his work. He will talk about his machine, and possibly operate it, outside work hours.

(d) A decrease in the frequency of interaction between a man and his machine entails a decline in the extent to which norms are clear.

(e) Men are more like the machines with which they interact frequently than like those with which they interact infrequently. One can see this, for instance, when operatives whose machines perform identically attribute differences in output to the machines, rather than to themselves; and each machine thus acquires a reputation. Other operatives expect the machine to conform to its reputation; and to some extent their own behaviour towards the machine seems designed to confirm the reputation. The 'good' machine may be carefully tended, and remain in good condition; the 'poor' machine may be treated roughly, adjusted by unskilled hands, and perform poorly eventually because it has been poorly maintained.

(f) The social ranking of men relative to machines within a factory will tend to carry over into their relationships outside the factory.

(g) The social ranking of the machines (in terms of expense, newness and performance) will tend to be reflected in the social ranking of the men who control and the men who operate them. New, expensive equipment brings prestige to those who control it. When automation means that a man has been put in charge of a machine—i.e. that he initiates their joint activity more often than does the machine —it raises his status. When automation puts a man in a position where the machine initiates most of their activity and determines the man's movements and pace of work, it lowers his status.

(h) So far as two persons both work on the same machine, the relationship between them will be one of constraint, and interaction will be kept to a minimum.

Where machines differ widely in expense and complexity, automation does not divide employees clearly into two classes, the controlling and the controlled; it introduces many small status differences

[66] *Ibid.*, pp. 92–114.

which are not simply the result of better performance by the more expensive machines.

The second part of Wirth's final proposition relates to the impact of the pecuniary nexus. This can most clearly be outlined in relation to the 'vocations'. The vocations differ from the other professions in a number of respects. Firstly, their vocational ideology stresses the goal of humanitarian service, based on devotion rather than material rewards. The activities of the nurse, for example, are viewed as valuable in themselves, and are assumed to need no justification in terms of economic rewards: one enters nursing because it is good, not because it is lucrative. Nor do they need justification through political action: the nurse is not expected to strike or lobby for higher salaries or better conditions. It is assumed that her superiors will recognize her worth and make sure that facilities and salaries are adequate. If reminders to the superiors are needed, they should be discreet and not public.

Secondly, in the vocational ideology it is legitimate to reject members of the profession on grounds other than academic inadequacy. Under this ideology, dedication is stressed more than intellect; and it is legitimate to discourage the person with outstanding technical knowledge but the 'wrong personality'. Training schemes for the vocations may include a long period of 'mortification'—floor-scrubbing for nurses, minutely detailed supervision for lady elementary school teachers. The major function of mortification is not to train the recruit technically, but apparently to weed out those who lack the dedication to complete the training course.

Thirdly, the vocational ideology tends to be peculiar to groups with a certain position in society. Most of its adherents perform personal services for all groups in the society, and not simply for those who can pay most. They perform these services as members of large organizations, not through private practice or specialized firms. Their salaries are paid on the basis of general competence rather than individual brilliance or the wealth of the clients they attract. Social control of the practitioner's conduct is achieved through dedication to the profession, rather than through loss of clients.

Fourthly, the ideology is functional in assisting its adherents to adjust to their limited salaries and poor working conditions, by comparison with other professionals. It encourages them to formulate goals which focus on effective performance under difficult conditions, rather than on elaborate amenities and high salaries. The ideal represents a real challenge; and because it is stringent, it binds together those who are maintaining it.

This picture of the vocations as middle class groups with a particular type of working conditions who accept a particular ideology, does not imply that every nurse is a vocational, while every lawyer is a professional. There are likely to be consistencies, however, between working conditions and ideologies. As working conditions change, acceptance of the ideology is changing. The increasing bureaucratization which has accompanied growing state control has tended to undermine the vocational ideology in two important respects. Firstly, it has tended to increase the number of supervisory layers over members of the vocations, and hence complicated the internal channels through which the vocational member could formerly ensure that his requests for improvements were heard. The dissatisfied vocational may now be in a weaker power position as a result. Secondly, and perhaps more important, increasing central control has tended to mean more detailed rules and greater encroachment on the professional's area of autonomy. While this has probably encouraged uniformity of professional standards it has also left less scope for individual initiative. The regular complaints from teachers about the 'tyranny' of the syllabus for a regional or national examination is a response to reduced autonomy. These changing conditions have tended to induce vocationals to conventionalize their dissatisfactions into requests for higher salaries and fewer sub-professional duties; and to organize in more bureaucratic and also more militant professional associations.[67]

Conclusion: Some Functions of Bureaucracy
The manifest function of any bureaucracy is to carry out the policy for which it was established. This is a more complex assignment than might appear on the surface, since it normally implies the translation of abstract policy principles into concrete administrative acts. The bureaucracy thus functions, not merely to obey directives, but to define them more precisely than legislation can, and to devise means of attaining them as fully as possible. Administration necessarily involves a variety of compromises between ends and over means: maintaining a balance, for instance, between economical general oversight and scrupulous but expensive attention to all details; between personal service tailored to the client and impersonal justice; between slow, careful, expensive decision-making and quick, economical action which may be high-handed. Finally, conflicts of interest inevitably arise between the policy goals and the needs of the organization set up to achieve them. While many decisions may be referred back to the original policy-makers, delegation of authority implies that many decisions will be taken at lower levels, possibly without the policy-makers' knowledge. The manifest function of translating policy into action is much more positive than mere obedience to rules.

[67] Tropp, A. *The School Teachers*. Heinemann, 1957. In the U.S., high school teachers accept lobbying as a vital activity of their professional association: Zeigler, H. *The Political World of the High School Teacher*. Univ. of Oregon, 1966, pp. 55–57.

The process by which the translation is made is complex; we have looked at the possibility of 'sabotage', in which the original goals become altered or even reversed. The problem of spelling out concepts such as justice and equality into daily acts is by no means easy. One might suggest that it will be done successfully in so far as (a) justice can be reduced to a set of instructions, based on a precise definition; (b) the effects of justice and injustice are readily visible; and (c) injustice will lead to scandal directed against the department.

The first latent function of bureaucracy is self-preservation. Selznick argues that 'It is of the essence of (bureaucratic behaviour) that action formally undertaken for substantive (i.e. policy) goals be weighed and transformed in terms of its consequences for the position of the officialdom.'[68] Lipset's *Agrarian Socialism*[69] documented the resistance by a conservative civil service to radical policy changes which threatened both the officials' posts and the set of rules by which they worked. A major re-writing or re-interpretation of the rules is itself no small undertaking: assuring that the habits of the many officials change correspondingly, and that clarifications are made to settle disputes, is a massive task. While bureaucracies tend to be regarded as conservative forces, it is important to avoid stereotyping them. Some of the policy changes may indeed be impracticable, in the sense that a bureaucratic structure is quite inappropriate for applying them. Many commercial bureaucracies initiate change; and government departments may have the same effect—when for example they take legislation more literally and administer it more enthusiastically than had been intended or when they press for a more aggressive foreign policy and subsequently expand to meet their new commitments.

The second latent function of bureaucracy is the encouragement of rationality and differentiation throughout society. Those who are united under impartial rules are more likely to unite with others who are governed by the same rules. Unionization has been spreading into middle class occupational groups where it was once unthinkable. 'Household' work is being sub-divided and performed by commercial firms with special equipment: the window-cleaning firm, the vacuum chimney sweeping company and the food vending firm are beginning to replace the independent artisans, the part-time lady workers, and the few remaining servants. In some American cities one can even hire a house-cleaning firm when the Negro maid resigns.

Equally striking, perhaps, is the emphasis on rationality and merit in situations where public scrutiny was formally unthinkable. Within the last ten years, for instance, a government committee has been

---

[68] Selznick, *op. cit.*, p. 32.
[69] Lipset, S. M. *Agrarian Socialism*. U. Calif. P., 1950.

investigating university teaching methods.[70] Although it was limited to factual matters, and revealed few scandals, the idea that professors should be skilled in lecturing is relatively new; and except in education departments, the lecturer who is trained to lecture well remains exceptional. Again, in the late fifties *The Times* published a feature article which purported to show that the forty-three Anglican bishops were rationally, if not democratically appointed; and in fact represented a reasonable cross-section of the population. One had been the son of a manual worker, one had been in a concentration camp, less than forty were the sons of clergy, and slightly less than forty were Oxbridge men. The point here is not the pathetic inaccuracy of the claim to representativeness, but the fact that *The Times* and the author should conceive that it was necessary or desirable to try to justify church appointments *on the grounds of representativeness*.

The preceding sections have presented a picture of bureaucracy, within the general framework of Wirth's theory of urban social organization. While some aspects of the theory have immediate validity, others have needed qualification and at times drastic revision. Three basic points of criticism have emerged; these parallel the criticisms made earlier. Firstly, Wirth over-simplifies by stressing only formal relationships; in so doing he ignores the complementary nature of primary and secondary group ties, as a result of which one never finds one type in isolation. Secondly, he is inclined to under-emphasize integration based on widespread acceptance of certain values and institutions; while this acceptance has been weakened by increasing differentiation and specialization, it has certainly not disappeared. Wirth was aware of this to some extent, and frequently stressed the importance of analysing the assumptions which members of a group take for granted. These widely accepted beliefs are important both in legitimating the bureaucracies and in motivating people to conform to many of the practices which they encourage. Thirdly, Wirth's theory ranges far in analysing urban organizations; its very complexity tends to obscure the emergence of a viable classificatory scheme for bureaucracies. So many elements are involved in the analysis that one is left uncertain which are essential and which are relatively trivial. We have tried in the discussion to clarify this, but considerably more research and thought is needed before we have a fully satisfactory theory of bureaucratic organization.

[70] The Committee on University Teaching Methods of the University Grants Committee.

6
Wirth's Theory of Urbanism: An Overall Evaluation

THIS final chapter is divided into three sections. The short first section is concerned with some of the omissions of the present book. The second makes a brief general evaluation of each of the twelve propositions derived from 'Urbanism as a Way of Life'. The final section reviews some of the elements of the city which other writers have regarded as crucial, but which are minor elements in Wirth's theory.

To place the evaluation in perspective, it may be useful to note briefly the respects in which our examination has been relatively thorough, and those in which little has been attempted. A short book of this nature can include all aspects of the city only by being consistently glib and curt. The analysis began with the premise that the city could be understood only in its social context. We therefore accepted Sjoberg's fundamental distinction between pre-industrial cities (which are found in feudal but not in folk societies) and cities in urban-industrial societies. The second chapter gave a general account of some of the conditions for the development of cities as a form of social organization, and of pre-industrial cities in particular. The three following chapters concentrated on cities in urban-industrial societies: the third examined Wirth's propositions in relation to social behaviour in the city. The fourth chapter was particularly concerned with the ecological aspects of the theory, and the processes through which men distribute themselves over the land and use its resources. In the fifth chapter, we re-examined the propositions in relation to bureaucracy. The data of the third chapter had suggested that Wirth's theory was very plausible as an account of relations between two city-dwellers chosen at random, but a much weaker account of relations between two members of a primary group within the city. To judge its usefulness, it was therefore desirable to apply it to intermediate situations, where people had regular but not necessarily intimate contact. Pure bureaucracy is very appropriate for this purpose, both because it has recently attracted widespread research interest, and because it is peculiarly urban.

Perhaps the two most significant omissions, from a theoretical point of view, are a separate account of social processes within urban-industrial cities, and an analysis of processes of transition from folk to feudal and from feudal to urban-industrial societies—or at least from pre-industrial to industrial cities. Social processes have generally been treated incidentally, as reference has been made to mechanisms of social control, institutions which functioned in assimilating newcomers, and the necessity for certain values and practices which integrated sub-groups. The omission of the process of urbanization was influenced by the appearance of Reissman's *The Urban Process* and Breese's *Urbanization in Newly Developing Countries*. The process of transition is still only poorly understood, though increasing attention is being paid to urban development in non-western nations. While Reissman offers an analysis in terms of four types of development— urban growth, industrialization, nationalism and the emergence of a middle class—and obtains rough measures for these, he is inclined to assert without real evidence that development is smoother when all four proceed at roughly the same pace. Conversely, in his view, pathologies occur where development is very uneven. Although he is apt to make unwarranted generalizations, Reissman's outline marks a useful step forward. It remains to examine the interrelations of his four types of growth, and then to test both his general scheme and his claims about pathological development.

The other omissions from this book fall into three groups. Firstly, no attempt has been made to incorporate a history of urbanization. Lampard and others[1] have collected material on this topic, and Max Weber[2] made an elaborate comparative analysis of the social structure of ancient and mediaeval cities in Europe and the Mediterranean region. Dickinson[3] has written at length on the more recent history of the West European city; and there are extensive histories of individual cities such as Gill and Briggs' history of Birmingham.[4] Since only a sketchy outline could have been presented here, it seemed more appropriate to refer the interested reader to these other sources.

Secondly, most of the 'substantive fields' have received scant attention. One could have devoted two pages each to urban politics, urban education, urban religion, the urban family, and written briefly on each major institution; but this would have been a superficial and often repetitive endeavour. Some of these institutions are being

[1] Lampard, E. E. 'The History of Cities in the Economically Advanced Areas'. *Econ. Develop. & Cultural Change*, vol. 3, 1955; Lampard E. E. 'Historical Aspects of Urbanization', in Hauser, P. M. and Schnore, L. F. (eds.) *The Study of Urbanization*. Wiley, 1965; Glaab, C. N. *The City: A Documentary History*. U. Wisc. P., 1958.
[2] Weber, M. *The City*. (Eng. trans.) Free Press, 1960.
[3] Dickinson, R. E. *The West European City*. Routledge and Kegan Paul, 1951.
[4] Gill, J. and Briggs, A. *A History of Birmingham*. O.U.P., 1951.

covered by other volumes in this series; and in so far as they are bureaucratically organized, chapter 5 offers a variety of general observations about them.

Thirdly, no special mention has been made of 'urban problems' or their solution. The problem of integrating diverse sub-groups has been raised; but we have not tried to chronicle the conventional problems of slums, congestion and delinquency and so on, nor to advocate our own favourite solutions. It is an open question whether these problems are soluble within the present social context; their solution may well require more radical changes in the social structure than most of us are willing to countenance. If, for example, male delinquency is in large part attributable to lack of opportunities for high-status positions, there would appear to be only one solution: more effective means of persuading males to be satisfied with middle and low status positions. While better schools may lessen the disadvantages which working class children face in educational competition, they can only have marginal effects on the proportion of high status positions available in the society. Similarly, as we mentioned briefly in chapter 4, the cost of resisting deconcentration is extremely high, not only in financial terms, but also in terms of the social changes which would be needed to make it effective.

I. WIRTH'S PROPOSITIONS: A SUMMARY

Each of Wirth's propositions has been examined in three contexts: that of the pre-industrial city, as described by Sjoberg; that of the urban-industrial society in general; and that of the bureaucracy. The next task of this chapter is to bring these examination results together, pointing to any conclusions which emerge consistently. It is not very satisfactory to summarize the analysis of each proposition in a few words; but where tentative generalizations appear feasible, they will be made.

1. Relatively weak bonds among co-residents; formal social control; physical separation of diverse sub-groups

The evidence on common bonds has generally been supportive of Wirth's propositions, with certain modifications. In pre-industrial cities, common bonds tended to be either very strong or quite weak. Occupational, religious, ethnic, kinship and social status groups frequently coincided: within these, and particularly within the elite, bonds were strong; but between members of different groups they were weak. In urban-industrial cities, they were strong only between intimates and in a few areas where occupational and kinship groups had survived and geographical mobility was low. Size, density and heterogeneity do not explain these 'survivals', though geographical immobility may.[5] Among co-residents and among co-workers they were frequently weak, although informal groups of moderate strength formed where people had sufficient common interests to work together in dealing with organizational strains and in clarifying role expectations.

Formal social control was indeed found to be important, especially in the government of relationships between different sub-groups, and between groups which were physically separated. Within an organization, however, formal control might be less important than a series of overlapping networks of informal control, each covering persons at only one or two levels. Formal control was likely to be most important at the lower levels of the bureaucracy, and of decreasing importance as one proceeded up the hierarchy.

Physical separation of clearly different sub-groups was clearest in the pre-industrial cities. In industrial cities, separation is real enough, but the boundaries between groups are blurred and subject to frequent change as invasions and successions take place. The complexity of the system of differentiation, and the limited currency of many of the status symbols, also means that the stratification system is blurred; indeed it would be more profitable to refer to a number of different status systems which overlap to some extent. Similarly, within bureaucracies there is an elaborate system of differentiation which blurs simple status comparisons. Here, in addition, physical separation may be deliberately restricted as a step in improving communication.

2. Difficulty of knowing all others personally; impersonality, superficiality, etc.; relationships treated as means to ends

By definition, increasing size makes it more difficult to know all others personally. This may not be crucial, however, if it is still possible to know all *significant* others. It may be important for the bureaucrat to know personally all those with whom he needs to interact regularly; but it may be of little importance whether he knows personally those persons with whom he rarely needs to communicate. In a bureaucracy, for instance, personal relationships may be established by persons at each level with their immediate superiors and subordinates; such a chain could extend throughout the organization and produce some of the results which follow when everyone knows all others personally. Personal relationships could also be significant if a social group became numerous, but if its members could still check each other's social acceptability by claiming mutual friends, acquaintances and relatives.

The 'syndrome' of impersonal, segmentary, transitory and superficial relationships persists only under certain conditions, even if it is

[5] Gans, H. 'Urbanism and Suburbanism as Ways of Life', in Rose, P. I. (ed.) *The Study of Society*. Random House, 1966, p. 311.

viewed as the organization's ideal. Trading relationships in pre-industrial cities did not fit this picture, even where little intimacy was involved. Impersonality and segmentation tend to break down where a person plays a number of roles toward another. It is also unlikely to persist where the roles are played over a considerable period, and where at the same time each party could exercise significant sanctions over the others' behaviour. Consequently some groups are insulated from the effects of size, density and heterogeneity by their structure and cultural patterns.[6]

The treating of relationships as means to ends is an imprecise phrase. We take it to mean either exploitation, a term which Wirth uses explicitly in a later proposition, or the practice of regarding the rules as more important than the other parties to the relationship. Although not concerned with analysing motives here, exploitation is clearly possible in both primary and secondary relationships. Perhaps it would be fair to say that primary and secondary relationships offer opportunities for different types of exploitation for different periods of time. Sustained exploitation may be more difficult to exert in a secondary relationship, and at times primary relationships may be avoided for this reason. Treating relationships as means to ends, in the sense now of regarding the rules as more important than the other persons, was found to be expected from lower-level bureaucrats, for the purpose of impartial administration. This becomes more difficult when prolonged contact occurs, or where each party is in a position to exercise effective sanctions over the others.

3. Division of labour, with relationships as means to ends; dominance of the large firm; codes of ethics

The division of labour is clearly a significant feature of pre-industrial and urban-industrial societies. While specialization influences efficiency, it may stop far short of, or continue far beyond, the point of greatest efficiency. The pre-industrial city has division of labour by product, but rarely by process. Large firms have very rarely been found in pre-industrial cities; division of labour by process occurs only in the production of luxury goods for the small elite market. In the urban-industrial society, large organizations are dominant in business, government and many other fields; yet small organizations have survived by dealing in either extremely specialized goods and services, or durables where purchases are infrequent, careful comparisons are common, consumers are ignorant, and idiosyncracies of styling become important.

Codes of ethics develop in formal groups, occupational groups and informal groups; their general unifying function is weakened where there is an elaborate division of labour, which narrows the area of

[6] *Ibid.*, p. 309.

common interests shared by members of a group. There may be no substantial code of ethics which applies to all members of the organization or city.

4. *Division of labour grows with the market; extreme specialization and interdependence*
Although the first half of the proposition can be supported in general we lack detailed knowledge of the relationship between growth in the division of labour and growth in the market, and about the conditions under which it is closest. Some goods and services offer much greater opportunity for specialization than others; and the curves relating market size to profitability and increases in the division of labour may be far from smooth.

Extreme specialization and interdependence is unstable, in the sense that a single, small uncooperative group may bring the work of the entire organization to a halt. Particular cities may also be unstable, in the sense that they have weak defences against destruction.[7] At the same time, *as forms of social organization*, the city and the bureaucracy are extremely stable. If one is destroyed, other similarly structured organizations usually replace it.[8] Similarly, Dahl[9] has noted the extreme stability of democratic two-party government in New Haven, Conn., and attributes this in part to the fact that although many specialized interest groups may be dissatisfied with it as a form of organization, their very diversity makes it unlikely that they will find an alternative which is more widely acceptable than the present form.

5. *Impossible to assemble all residents together; indirect communication by mass media and special interest groups*
Again, the first half of the proposition is usually self-evident. The result, in the pre-industrial city, tends to be erratic communication since there are no mass media. In urban-industrial societies, formal communications tend to be mediated by primary groups, which evaluate and reinterpret them in many cases. While the primary group may thus insulate its members in the short run, it would be much more difficult to establish whether it can consistently protect its members

[7] Pre-industrial cities were frequently destroyed in wars; and in a few instances were tied to royal courts, which moved at the pleasure of the king: Bascom, W. R. 'The Urban African and his World'. *Cahiers d'Etudes Africaines*, vol. 4, 1963. Their reconstruction then depended on the advantages of their sites. As cities become larger, the chances of their destruction or abandonment appear to lessen, and their permanence increases.
[8] Similarly, Western observers have frequently misjudged the stability of dictatorship. Most individual dictatorships are indeed unstable; but when one dictator is overthrown, his successor is usually equally dictatorial. Hence dictatorship *as an institution* may be extremely stable, even though individual dictatorships fall very regularly.
[9] Dahl, R. A. *Who Governs?* Yale U.P., 1961, pp. 311-325.

against the longer-term effects of constant exposure to a single point of view, especially on subjects which are not highly salient to group members. Special interest groups may therefore be most effective when they face little opposition. Increasing size has various effects on the force of an interest group's communication. On the one hand, it increases the potential pressure that can be brought to bear, since the group becomes more representative of the entire population; on the other, it decreases the probability that the whole membership can be organized to take concerted action, should the threat of action itself prove insufficient.

6. *Differentiation and specialization increase as density grows*
This is a very plausible result of increased density, though there is little evidence that it is inevitable. Efforts to increase the size of the market, and approximation to the pattern of a total institution, may be alternative solutions under some circumstances to the changes necessitated by increasing density.

7. *Close physical but weak social bonds; readily perceptible symbols*
One normally associates close physical with close social bonds, and distant physical with weak social bonds. The maintenance of close physical and weak social bonds is made much easier if interaction is mainly formal and is based on readily perceptible symbols and rituals. Bureaucracies may stress physical proximity, or weak common bonds, but it is difficult to maintain a balance which incorporates both. Where both are maintained, but the common goals are few and not very salient, mob activity is likely to result. Where contacts are mainly superficial, symbols are an important basis for role recognition. Where they are associated with status differences, they are likely to become the objects of competition between groups with differing prestige.

8. *Economic competition determines land use; residential desirability is complex; segregation of commercial and residential land use*
Land use competition is much more complex than Wirth suggests: it involves political and value decisions at many points. This is particularly true in pre-industrial cities, where political and religious criteria tend to be dominant. Many buildings are erected for status as well as economic purposes; others for leisure pursuits to which no economic value is attached. This should not suggest a simple non-economic explanation; but rather a complex explanation in which economic competition is frequently the form in which other types of competition find expression.

Residential desirability tends to be much simpler in pre-industrial cities: the central site is desirable from most points of view, and is crucial to the solidarity of the elite. Desirability decreases with distance from the centre. As transportation improves, desirable areas tend to move outward; although some central sites are occupied by luxury flats, the affluent tend to move to more secluded areas, preferably with ready access to their favourite leisure pursuits. Those who remain near the centre tend to form enclaves, in buildings which are carefully protected from 'intruders'.

The segregation of commercial and residential land use is found only in urban-industrial societies. Sometimes it is rationalized by planners who enforce segregation, for the residences of all classes if possible. In developing and pre-industrial cities, it is rarely found, unless the city has a western colonial tradition. Where there are few large organizations, most producers live at their workplace; since there is little family life apart from the activities of their workplace, this does not require a special building. At most, they may work in one room and sleep in another.

9. Competition and mutual exploitation rather than cooperation at work; life at a fast pace; orderly meaningful routines
Competition and mutual exploitation among co-workers is formally stressed much less than cooperation and loyalty. Pre-industrial relationships between seller and buyer are surrounded by time-consuming rituals; those between master and apprentice by elaborate codes of obligations. While these do not prevent exploitation, they restrict it within conventional limits, and in this way limit the resentment it can arouse. We suggested in the third chapter that exploitation was clear only when the gains of one party were exactly equivalent to the losses of all the others combined; where interaction brought benefits to both simultaneously, it is difficult to judge whether exploitation took place. Indeed, the problems of defining exploitation increase rapidly as the number of affected parties grows. In urban-industrial societies, the formal and informal rules presume cooperation while frequently encouraging competition. Rules may be devised for restraining and determining the outcome of competition; but wherever promotion depends on merit rather than on seniority, competition is tacitly encouraged.

Little has been said about the constant stimuli of living at a fast pace. The impact of frequency depends on the degree of familiarity with the stimuli: where these are strange, their frequency is much more closely associated with perception of a fast or slow pace. Wirth's description is in many ways accurate, in that the pressure of work on the top officials may grow proportionally with the size of their organizations and with the complexity of the problems they face.

Orderly meaningful routines are of the essence of bureaucracy, though there may be regular conflict between the official routines and those of the informal group. Routine is found much more at the lower than at the upper status levels, however. In the pre-industrial city, seasonal routines tend to be much more orderly than daily routines. Religious celebrations are ascribed to particular days; but the time when the merchant opens or closes his shop may be quite unpredictable.

10. Simple class distinctions break down as the division of labour becomes complex

It was not clear from Sjoberg how far the pre-industrial city retained a simple class structure. The broad distinctions between the elite, the lower class and the outcastes were clear; but there may well have been considerable complexity within each. In urban-industrial societies, there are a number of differing bases for the attribution of status; and these do not always coincide closely. Even when there is an official attempt to impose a single salary structure, this may appear artificial and be a considerable source of dissatisfaction. Complexity allows much more sideways mobility, whereas a simple system permits only movement upwards or downwards.

11. Group loyalties conflict; geographical and social mobility; sophistication

Conflicts of loyalty are much less frequent in pre-industrial cities, and mobility is limited; consequently one would expect less exposure to differing viewpoints and less sophistication. The elite in particular avoids contact with the other classes; they would therefore be expected to become sophisticated only in interpersonal relationships. Conflicts of loyalty are not peculiar to urban-industrial societies; but in folk and feudal societies those who play roles which conflict also play roles toward each other which stress cooperation. Conflicts of group loyalties become more easily conventionalized because the groups to which one belongs are more easily arranged into a hierarchy. The vaguer nature of the hierarchy in urban-industrial societies accounts for the special significance of conflicting loyalties in them.

Geographical mobility appears to be higher in industrial cities than in rural areas, though this may be partly an illusion. Rural depopulation alone would give the impression that the city population was more mobile because it always contained proportionately more recent arrivals. There are no reliable data on mobility in pre-industrial cities; though studies in West Africa suggest that there may be very large seasonal movements into and out of the cities.

Sophistication in Wirth's sense is sometimes found in pre-industrial cities, but many of these have confirmed and elaborated rather than challenging traditional religious and social beliefs. Sophistication is likely to be much more frequent in urban-industrial societies, where no single scale of values is so generally accepted, and there is more opportunity and encouragement for thinking out one's own scale.

12. Levelling influence of mass production; pecuniary nexus
This applies only weakly in pre-industrial cities: not only is there little mass production, but trade is treated with contempt as an unworthy pursuit. Mass production is much more significant in urban-industrial societies, where its influence is readily visible. It may of course be more apparent than real: the opportunities for variety in material, for example, are much greater now than ever before. Within the organization, mass production has a levelling effect only in so far as it reduces all jobs to labouring status. Differences between machines, and in the relations between man and machine, create new distinctions.

The pecuniary nexus, similarly, tends to be much less apparent in pre-industrial cities. In urban-industrial societies, it is very readily apparent. This can be seen most obviously in the widespread belief that people can be attracted into almost any job, provided only that the pay is high enough; and that any service for which one can persuade people to pay needs no further legitimation. The influence of the pecuniary nexus can be exaggerated, however: while it is conventional to express many types of dissatisfaction as demands for higher wages, there is ample evidence that monetary changes do not solve the underlying problems.

II. OMISSIONS FROM WIRTH'S THEORY

This final chapter concludes with an assessment of some of the additional elements which one would wish to see in a theory of urban social organization. Obviously Wirth's theory has proved much more applicable to industrial than to pre-industrial cities. In the former case, some of his propositions have required serious qualifications; but in the latter, some have scarcely applied at all. The inferences about the social patterns which are needed to integrate the city have depended on inappropriate concealed assumptions. The most glaring shortcomings of Wirth's theory in this context have been the concealed assumptions that economic rationality is highly valued and that large pure bureaucracies are feasible. This is not the place for a detailed analysis of their prerequisites: we note only Wirth's implicit assumption that the prerequisites were available in pre-industrial cities. Where they are missing, large scale organization is likely to take the form of patrimonial bureaucracy. Wirth's analysis also raises a question which Sjoberg does not treat satisfactorily in *The Pre-Industrial City*: if there are clear sub-groups within the pre-industrial city, which differ in many respects and which interact only irregularly, what mechanisms integrate the city population? A traditional religion and value system are often important; economic convenience might also be significant, as Wirth himself noted; so no doubt are the political and military power of the elite. The propositions which relate to the integration of diverse groups thus have some applicability to the pre-industrial city; although the range of solutions which Wirth envisaged is sometimes too limited to account even for western cities.

A second valid criticism of Wirth relates to his account of the functions of primary groups. He was fully aware, from his studies of ghetto dwellers, that primary groups existed among city dwellers; but his theory was unable to assimilate the idea that they had essential functions in urban society. The ghetto has functions other than the assimilation and protection of its members. Primary groups and relationships are sometimes survivals from a traditional rural culture; but at others they are integral aspects of urban society, and are highly functional in dealing with the strains in urban life, and in filling in gaps in its structure. They may, indeed, be supplemented by 'quasi-primary' groups, especially in the suburbs.[10] It is therefore legitimate to criticize Wirth for making no reference to the crucial part played by the latter in urban society.

The third criticism which has been legitimately levelled against Wirth is that he pays only limited attention to the third of his criteria, heterogeneity. He is acutely aware that heterogeneity is a potential source of conflict and malintegration; and he is aware that in avoiding this potential problem, people tend to surround themselves with like-minded others. Yet he does not stress that this may be a powerful source of integration; although the congregation of like-minded neighbours in an area may be largely unintentional, the possibility of finding a considerable number of persons who share one's own values may be a very real attraction of the city. Unless its population is highly selective, the small town is unlikely to offer such a possibility to its recent arrivals. The attraction of the presence of a limited number of persons sharing one's particular interests may more than offset the reluctance to live near much larger numbers who have very different opinions.

Wirth also tends to underestimate the possibility that the heterogeneity of the city population makes simple generalizations invalid. The effects of size and density may depend heavily on the composition and values of the population; and only gross deductions can be made about their impact on social patterns. From this argument, two conclusions would follow. Firstly, the propositions may need different modifications to apply to different sections of the population. The initial deductions—for example, that as the group increases in size the possibility that everyone knows everyone else personally will decrease—are usually very plausible; but the necessary conditions for

[10] Gans, *op. cit.*, p. 312; Form, W. H. *et al.* 'The Compatibility of Alternative Approaches to the Delimitation of Urban Sub-Areas'. *Amer. Sociol. Rev.*, vol. 19, 1954.

integration (parts (b) and (c) of each proposition) are by no means the only ones possible. One illustration will perhaps suffice. The mass media and special interest groups are alternatives to direct personal communication, but they are not the only alternatives. An equally feasible pattern in bureaucracy is an interlocking series of informal groups, each spanning two or three levels in the hierarchy, through which communications pass. This represents an alternative method of integration when the group is too large for all members to communicate personally with all others.

The second conclusion which Wirth might have drawn from the heterogeneity of the city is more complex. Mann points to it with the paradox that a small, relatively sparsely populated London commuter village may exhibit more urban behaviour patterns than a larger, denser town. The suburb is outside the London conurbation; the town is inside the West Riding conurbation. The differences in the sizes of the two conurbations hardly constitutes a sufficient explanation, since the paradox apparently flatly denies many of Wirth's propositions. It is not simply a reflection of the distinction between urban and urbane. The commuters may also have been more urbane than the town-dwellers; but this does not resolve the paradox. The problem goes deeper: the differences between the moderate-sized town and the commuter suburb are poorly predicted by Wirth's propositions; but this is not because rural traditions survive in the commuter suburb, nor because it is less integrated into British industrial life.

To understand the paradox, we must realize that Wirth's conclusions about the sources of integration in urban life could have been deduced from a definition in terms quite other than size, density and heterogeneity. This is not to argue that size, density and heterogeneity are trivial and unimportant; but that they are only one of several bases from which one could deduce the twelve propositions about urban life. We argued above that Wirth's deductions depended on concealed assumptions about the importance of rationality and the feasibility of large-scale organization. Wirth began by attempting to account for urban-rural differences, although this was rarely stated explicitly. In this context it was natural to assume that the primary features of the city were its obvious differences in size, density and heterogeneity from the country-side. As urban-rural differences diminish in urban-industrial societies, the significance of size, density and heterogeneity inevitably appears to weaken. Some writers have tried to resolve this difficulty by distinguishing between urban societies which have both cities and urban-rural differences, and mass societies, which have metropolitan areas or conurbations and relatively few urban-rural differences. The distinction is one of degree rather than kind; Wirth's analysis would in this respect apply better to the urban than to the mass societies.

We are now ready to re-phrase this second conclusion from the arguments concerning the place which Wirth gives to heterogeneity. The key features of the city are not the inevitable, logical consequences of the size and density of its groupings, but of a complex combination of features: its technology and communications system, its values and social structure, its social and physical environment, for instance. One could deduce many features of urban-industrial society from assuming the primary of any one of these: a utopian beginning with the value of economic rationality might well conclude that precisely the same city characteristics were necessary for integration. In the same vein, Weber's argument in *The Protestant Ethic and the Spirit of Capitalism* could be mistakenly taken to support the view that urban-industrial society is in large measure the unintended consequence of the late mediaeval religious upheavals.

While not suggesting that Wirth believed size, density and heterogeneity to be the major *causes* of the form of social organization in urban-industrial society, his propositions raise the problems of determining cause and effect relationships. The inadequacy of the propositions in analysing pre-industrial societies leads one to conclude that they are insufficient as a causal explanation, whether or not Wirth intended them to serve as one.

In this conclusion, we have tried to show that: (a) if Wirth's propositions are treated as causal relationships, they represent an incomplete theory of urban development; and (b) it seems likely that his conditions for cohesion in urban-industrial society can be deduced from basic urban characteristics quite different from those which he chooses.[11] A criterion of economic rationality might produce very similar propositions, under which large size and high density might prove to be necessary conditions but not sufficient ones for social integration. This does not mean that Wirth's entire argument is circular and valueless. It is possible that, by beginning from different basic characteristics, one might obtain a model which would apply to all cities; but this is not too likely. Progress is more likely to come through a closer identification than Wirth made between the city and the society in which it is found; a classification of cities and societies together; and separate analyses for different types of cities, based on the recognition that the city as a type of social organization may arise for quite different reasons and be subject to quite different sets of conditions for survival in different societies.

[11] Gans, *op. cit.*, pp. 311–316, argues that residential instability, social class and life-cycle stage are critical; the first in explaining the 'social types' found near the city centre; the second and third in explaining many features of suburban life. While pointing to their importance, he does not develop an alternative set of propositions which could be compared with, or integrated with, Wirth's propositions.

BIBLIOGRAPHY

Bascomb, W. R., 'The Urban African and his World'. *Cahiers d'Etudes Africaines*, vol. 4, 1963.
Bogue, D. J., *The Structure of the Metropolitan Community*. Univ. Michigan, 1949.
Breese, G. E., *Urbanization in Newly Developing Countries*. Prentice-Hall, 1966.
Burgess, E. W. & Bogue, D. J. (eds.), *Contributions to Urban Sociology*. Chicago, 1964.
Caplow, T., *Principles of Organization*. Harcourt-Brace, 1964.
Dewey, R., 'The Rural-Urban Continuum'. *Amer. J. Sociol.*, vol. 66, 1960.
Dickinson, R. E., *The West European City*. Routledge & Kegan Paul, 1951.
——*City and Region*. Routledge & Kegan Paul, 1964.
Duncan, B., Sabagh, G., & Van Arsdol, M.D., 'Patterns of City Growth'. *Amer. J. Sociol.*, vol. 67, 1962.
Duncan, O. D. & Reiss, A. J., *Social Characteristics of Urban and Rural Communities, 1950*. Wiley, 1956.
Ericksen, E. G., *Urban Behavior*. Macmillan, 1954.
Etzioni, A. (ed.), *Complex Organizations*. Holt, 1961.
Firey, W., *Land Use in Central Boston*. Harvard, 1947.
——*Man, Mind and Land*. Free Press, 1960.
Gans, H., *The Urban Villagers*. Free Press, 1962.
——'Urbanism and Suburbanism as Ways of Life: A Re-Evaluation of Definitions', in Rose, P. I. (ed.), *The Study of Society*. Random House, 1966.
Gibbs, J. P. (ed.), *Urban Research Methods*. Van Nostrand, 1961.
Glaab, C. N., *The City; A Documentary History*. Univ. Wisconsin, 1958.
Hadden, J. K. & Borgatta, E. F., *American Cities*. Rand McNally, 1965.
Haig, R. M., 'Toward an Understanding of the Metropolis'. *Quart. J. Econ.*, vol. 40, 1926.
Halmos, P. (ed.), *The Development of Industrial Societies*. Sociological Review Monograph No. 8. Keele, 1964.
Hatt, P. K. & Reiss, A. J. (eds.), *Cities and Society*. Free Press, 1957.
Hauser, P. M. & Schnore, L. F., *The Study of Urbanization*. Wiley, 1965.
Jones, E., *Towns and Cities*. O.U.P., 1966.
Lampard, E. E., 'The History of Cities in Economically Advanced Areas'. *Econ. Develop. Cultural Change*, vol. 3, 1955.
Little, K. L., *West African Urbanization*. C.U.P., 1965.
Mann, P. H., *An Approach to Urban Sociology*. Routledge & Kegan Paul, 1965.
Mayer, R. F. & Kohn, C. F. (eds.), *Readings in Urban Geography*. Chicago, 1959.
Merton, R. K., *et al.* (eds.), *Reader in Bureaucracy*. Free Press, 1952.
Moser, C. A. & Scott, W., *British Towns*. Oliver & Boyd, 1961.
Reiss, A. J. (ed.), *Louis Wirth on Cities and Social Life*. Chicago, 1964.
Reissman, L., *The Urban Process*. Free Press, 1964.
Shevky, E. & Bell, W., *Social Area Analysis*. Stanford, 1955.
Sjoberg, G., *The Pre-Industrial City*. Free Press, 1960.
Stacey, M., *Tradition and Change*. O.U.P., 1960.
Williams, W. M., *The Sociology of an English Village: Gosforth*. Routledge & Kegan Paul, 1956.
Wirth, L., 'Urbanism as a Way of Life', *Amer. J. Sociol.*, vol. 44, 1938.

INDEX

Adaptation, 42–45
Age of Cities, see City, Types of
Areas, Residential, 18, 36, 37, 51, 80–81, 166–167
Areas, Social, 16, 18, 20, 25, 33–38, 49, 51–52, 62, 67, 70, 81, 86, 90–95, 96, 122, 167, 170
Assimilation, 67, 70, 98, 99, 101, 161, 170
Associations, Voluntary, 69, 71, 75, 77, 83, 84–85, 98, 100
Attainment, Goal, 42, 45
Automation, 154–156

Buildings, 145–146
Bureaucracy, 60, 82, 85, 86, 100, 113, 114–159, 162
Bureaucracy, Client Relationships, 130–131, 146; Functions of, 152, 157 159; Impartiality, 121–122, 125, 126–131, 138, 141, 148, 158, 164; Leadership, 118–119, 120, 121, 141 Line, 133, 151–153; Patrimonial, 114, 123, 133; Promotion in, 122, 124, 126–130, 136–137, 138, 14—148, 150; Recruitment, 127–127, 156, 159; Rules in, 121–124, 126, 133, 134, 138, 139–140, 144, 146, 147, 150, 158, 164, 167; Sabotage, 119–120, 139–140, 158; Sponsorship, 127, 138, 149; Staff, 133, 151–153; Supervisory Roles, 116, 118, 120, 122, 123, 132, 139, 140, 156–157
Business, Large-Scale, 17, 22, 28, 29, 31, 36, 37, 49, 50, 54, 55, 57, 58, 72, 73–74, 75–76, 130–131, 132–133, 136, 164, 169, 171

Capital, see Credit
Capitalism, 21, 28, 58, 59
Causal Relationships, see Relationships
Census, England & Wales, 26, 30, 70
Census, Tract, see Tract
Census, United States, 26, 30, 63, 88
Ceremonies and Rituals, 138, 142, 143–144, 148–149, 167
Centralization, 101, 102, 104, 107–109
Change, Social, 21–22, 26, 27, 40, 60, 82–83, 134

Cities, Classifications of, 21, 25–33, 47, 62, 75, 86–90, 172; Hierarchy of, see Hierarchy
City, Definition of, 16, 41, 62; Key Features of, see Size
City, Problems of, see Problems
City, Types of: Age Classifications, 21, 29, 87, 90; Administrative, 27, 30, 47, 57; Capitalist, 27, 90; Central Place, 27, 89; Colonial, 27, 30, 47; Cultural, 30; Economically Balanced, 29, 90; Feudal, 26, 27, 90; Generative, 27, 90; Heterogenetic, 26; Industrial, 21, 22, 23, 27, 28, 29, 31, 32, 33, 37, 39, 54, 55, 62–100, 101–113, 114–159, 162–169; Location Classifications, 21, 29, 89–90; Market, 27, 57; Metropolitan, 29; Orthogenetic, 26; Parasitic, 27, 90; Patrician, 26, 30, 89; Plebeian, 26, 30, 89; Postroad Stages, 30; Pre-Colonial, 27, 30; Pre-Industrial, 22, 23, 28, 29, 30, 32, 33, 39–55, 57, 58, 62, 76, 81, 83, 84, 90, 95, 134, 160, 162–169, 172; Primate, 57, 58; Primitive, 26; Prosperous, 29, 87, 90; Provincial Capitals, 30; Shrine, 30; Slave-Owning, 27, 90; Socialist, 27, 90; Temple, 30; Transport, 27, 89
Class Distinctions and Structure, 18, 22, 28, 29, 30, 31, 37, 51, 53, 64, 69, 70, 76, 82, 83, 87, 89, 149–150, 168
Class, Lower, 48, 50, 51, 52, 53, 68, 70
Class, Upper, see Elite
Communication, 17, 51, 52, 57, 66, 76–77, 101, 104, 107, 108, 124–125, 126, 127, 138–140, 141, 143, 146, 163, 165–166, 172
Competition, 18, 35, 43, 45, 47, 52, 59, 67, 69, 78, 80–83, 91, 104, 105, 111, 113, 116, 125, 132, 144–146, 147, 148–149, 152–153, 166, 167
Concentration, 57, 101–102, 103–107, 162
Conflict, Role, 83–84
Control, Social, 16, 48, 55, 66, 69, 101, 115, 117, 121–124, 126, 132, 137, 143, 152, 161, 163
Cooperation, 18, 23, 43, 47, 52, 59, 67, 69, 116–118, 120, 126, 127, 140, 142, 144, 146, 147, 149, 167

INDEX

Co-Residence, 16, 18, 48, 66–69, 162–163
Co-Workers, 18, 22, 37, 54–55, 71, 75, 81, 115–125, 128–130, 132, 133–134, 139, 141, 142, 148–149, 163, 167
Credit & Capital, 50, 54, 55, 56, 57, 60, 82

Decentralization, see Centralization
Deconcentration, see Concentration
Density, see Size
Differences, Urban-Rural, 23, 26, 38, 63, 65, 82, 83, 84, 85, 96, 99, 171
Differentiation, see Labour, Division of
Diffusion of Urban Ideas, see Ideas
Dominance, 30, 64, 73, 89, 102, 107, 133

Ecology, Social, 33, 91, 95, 101–113, 144, 146
Elite, 27, 28, 29, 39, 41, 43, 46–47, 48, 49, 50, 51, 52, 53, 55, 115, 162, 164, 166, 168, 170
Equilibrium, 17, 50, 74, 75, 134, 136–137, 149, 165
Ethics, Codes of, 17, 49–50, 53, 72, 74, 131, 133–134, 164–165
Exploitation, 18, 49, 52, 72, 81, 126, 146, 164, 167

Family & Kinship Groups, 22, 39, 50, 51, 52, 53, 54, 59, 68, 76, 84, 96–98, 99, 110, 162
Finance & Wholesaling, 27, 64, 66, 73, 89, 107
Firm, Large, see Business
Functions, Economic, of the City, 17, 21, 27, 30, 66, 75, 78, 86–90,

Ghettoes, see Areas, Social
Goods, Convenience, 74, 109; Luxury, 74, 109; Shopping, 109
Groups, Ethnic, 48, 52, 95; Kinship, see Family; Primary, 25, 38, 62, 68, 69, 72, 76–77, 95–100, 118–121, 132, 133–134, 139, 142, 144, 147, 159, 165–166, 170
Growth, Economic, 27, 31, 50, 55, 59, 60
Growth, Urban, 16, 20, 27, 31, 35, 57, 58, 64, 103–107, 161
Guilds, Occupational, 39, 46, 48, 49, 50, 51, 52, 53, 59, 74

Heterogeneity, see Size
Hierarchy, Urban, 57, 75, 90
Hinterland, 20, 27, 30, 31, 44, 89

Ideas, Urban, Diffusion of, 20, 99
Industrial Cities, see Cities
Industrialisation, Conditions for, 31, 39, 40, 41, 50, 55–61
Integration, Social, 17, 20, 21, 24, 42, 43, 46, 47, 72, 73, 74, 76, 101, 131, 133–134, 136, 137, 142, 159, 161, 169, 171
Institutions, Total, 115, 125, 141
Interests, Special, 17, 51, 76, 77, 138, 165, 166, 171
Invasion, 70, 101, 102, 109–112

Labour, Division of, 17, 18, 37, 44, 47, 49, 50, 53, 54, 55, 66, 67, 72–73, 74–76, 77–78, 83, 92, 97, 101, 107, 131, 134–137, 138, 140–141, 148, 149–150, 159, 164
Lampard, E. E., 27, 30, 57, 77, 161
Land Use, 18, 22, 29, 33–38, 51–52, 80–81, 90–95, 101–113, 144–146, 166–167
Land Use Theories: Concentric Zone Theory, 33, 34, 91–92, 94; Multiple Nuclei Theory, 34, 92, 94, 95; Sector Theory, 33, 34, 92, 94, 105
Large-Scale Organization, see Business
Location of Cities, see Cities

Maintenance, Pattern, 42, 45
Markets 17, 50, 51–52, 73, 74–76, 134–136, 165, 166
Mass Communication, see Communication
Mass Media, 17, 51, 54, 76, 138, 165–166, 171
Mass Production, Influence of, 19, 37, 53–54, 75, 78, 79, 85–86, 107, 154–156, 169
Mobility, 18, 48, 53, 58, 59, 66, 67, 71, 80, 83, 84–85, 89, 99, 116, 130, 151, 152–153, 162, 168
Multiple Nuclei Theory, see Land Use Theories

Nationalism, 31, 59, 161
Nexus, Pecuniary, 19, 53–54, 59, 73, 85, 154, 156–157, 169
Norms, Property, 58

176   INDEX

Others, Significant, 163
Outcastes, 48, 52, 53, 168

Pace of Life, Fast, 18, 52, 81, 82, 146–147, 167
Planning, Urban, 80, 81, 106–107, 112, 167
Pre-Industrial Citites, see Cities
Primary Groups, see Groups
Problems, Social, 22, 67, 105, 162
Processes, Ecological, 70, 101–113
Processes, Social, 62, 70, 101, 161
Production, see Mass Production
Professions, 17, 73, 74, 85, 127, 133, 138, 151–152, 156–157, 158; Vocational, 156–157
Property, see Norms
Proposition 1: 16, 48–49, 66–70, 115–125, 162–163; Proposition 2: 16–17, 49, 51, 70–72, 125–131, 163–164; Proposition 3: 17, 49–50, 72–74, 131–134, 164–165; Proposition 4: 17, 50, 74–76, 134–138, 165; Proposition 5: 17, 50–51, 76–77, 138–140, 165–166; Proposition 6: 17, 51, 77–78, 140–141, 166; Proposition 7: 17–18, 51, 78–80, 141–144, 166; Proposition 8: 18, 51–52, 80–81, 144–146, 167–168; Proposition 9: 18, 52–53, 81–83, 146–149, 168–169; Proposition 10: 18, 53, 83, 149–150, 168; Proposition 11: 18, 53, 83–85, 151–154, 168; Proposition 12: 18–19, 53–54, 85–86, 154–157, 169

Region, Metropolitan, 20, 27, 32, 66, 102, 105, 107, 108
Relationships, Causal, 21, 41, 172; Social, 16, 17, 18, 19, 22, 23, 24, 33, 37, 49, 52, 70–72, 78–79, 116, 117–118, 122, 125–132, 133, 136, 141–142, 163–164, 166; Trading, 49, 52, 61, 72, 74, 164, 167; with Clients, see Bureaucracy
Residential Areas, see Areas
Retreat, see Invasion
Rituals, see Ceremonies
Routines, 18, 52, 55, 60, 81, 82, 139, 146, 167–168
Rural-Urban Differences, see Differences

Sector Theory, see Land Use Theories
Segregation, 16, 48–49, 51, 66, 69, 80–81, 94, 95, 101, 102, 112, 115, 124–125, 144, 149, 162, 163, 166–167
Size, Density & Heterogeneity as Key Features of Cities, 15, 16, 17, 18, 20, 21, 24, 26, 29, 32, 36, 37, 41, 42, 46, 47, 51, 52, 55, 63–66, 69, 72–73, 77–78, 88, 89, 95, 133, 140–141, 162, 163, 166, 170, 171, 172
Socialization, 153–154
Society, Feudal, 28, 31, 61, 62, 63, 79, 160, 161, 168; Folk, 28, 39, 79, 98, 160, 161, 168
Sophistication, 17, 18, 53, 85, 151, 154, 168
Special Interests, see Interests
Specialization, see Labour, Division of
Stability, 55, 136; see also Integration
Standardization, 23, 49, 54, 74, 135
Structure, Social, 20, 21, 24, 27, 37, 39, 40, 41, 47, 48, 55, 62, 92, 161, 172
Succession, Ecological, 70, 101, 102–103, 109, 163
Surplus, Agricultural, 28, 31, 43–45, 46, 47, 57, 58–59
Symbols, 18, 19, 37, 51, 69, 78, 79–80, 126, 127, 141, 143–144, 163, 166

Technology, 28, 29, 31, 39, 43, 50, 54, 55, 56, 75, 87, 172
Tract, Census, 70, 93, 94
Trade, see Relationships, Trading
Transportation, 27, 34, 36, 44, 51, 52, 54, 57, 65, 66, 89, 95, 104, 105–106, 108, 145, 166

Urban-Rural Differences, see Differences
Urbanization, 20, 30, 31, 36, 37, 40, 54, 57, 58, 62, 75, 86, 91, 94, 95, 96, 97, 161; Conditions for, 43–47; History of, 161
Urbanism, 15, 19, 20, 37
Use of Land, see Land Use

Wholesaling, see Finance
Withdrawal, see Succession

Zone Theory, see Land Use Theories